Provocative Plastics

"The 19th-century inventors of plastic saw it as an alternative to precious materials such as tortoise shell and ivory, and in the 20th century it transformed design and our material world with a plethora of life-changing products. It was seen as a miraculous substance that fuelled remarkable innovation. Today its ubiquity and cheapness is cause for environmental concern: many of the 5 trillion plastic bags used each year clog up the world's oceans.

This book, in taking a considered, long view over plastic's more than 150-year history, shows how it has changed our planet for better and worse. It illuminates the cultural contribution and promise of plastics, as well as its throw-away and polluting characteristics. We are at a tipping point in our appreciation of this most malleable of materials, and this collection is provocative in that it suggests that, if broached with ingenuity and creativity, and used and disposed of responsibly, plastics can be part of the solution to the many problems it causes."
—Dr Tristram Hunt, *Director, Victoria and Albert Museum*

"Provocative Plastics sidesteps the over simplistic idea that plastics are either 'good' or 'bad' to provide a thorough account of plastics' continued importance to today's world. It is a must-read for those who feel they need to know more about these controversial materials before they decide what they think about them."
—Dr Penny Sparke, *Professor of Design History and Director, Modern Interiors Research Centre, Kingston University, London*

Susan Lambert
Editor

Provocative Plastics

Their Value in Design and Material Culture

palgrave
macmillan

Editor
Susan Lambert
Museum of Design in Plastics
Arts University Bournemouth
Poole, UK

ISBN 978-3-030-55881-9   ISBN 978-3-030-55882-6   (eBook)
https://doi.org/10.1007/978-3-030-55882-6

This Palgrave Macmillan imprint is published by the registered company Springer Nature Switzerland AG.
The registered company address is: Gewerbestrasse 11, 6330 Cham, Switzerland

*In Memory
of
Mark Suggitt,
A Remarkable Curator
1956–2019*

FOREWORD

On a rainy sunshine day in 2015, I was on the podium in the lecture theatre at the Arts University Bournemouth—a little nervous as always and excited. I was about to give the keynote at the *Provocative Plastics: Design in Plastics from the Practical to the Philosophical* conference, on which this book is based. I was pleased to participate because my passion is design and how design can make the world more sustainable and a better place for us all to inhabit.

To this end, I founded and was CEO of INDEX: Design to Improve Life—the world's biggest design award with a sole focus on sustainable design.

In this role, I was struck by the extent to which plastics, the materials that have become increasingly abhorred, were also those that featured predominantly in the manufacture of INDEX award winners. Plastics are now part of most of our activities and, indeed, are the materials that make many activities possible. Even single-use plastics have their place. As I write, during the COVID-19 pandemic, I am even more aware than usual of the value of plastics as an enabler in medicine and especially single-use plastics in terms of personal protection equipment.

None of this means plastics are not a problem. They are, as demonstrated by how they have and are contaminating large areas of land and sea. However, plastics are inanimate entities and have no ability of self-action. Mankind has created this problem and must find appropriate solutions. Proper disposal is vital and that is in our hands.

Arguably, plastics contain the solution as well as the problem. They play a key role in four of the six 2019 INDEX prizewinners, demonstrating in these different applications their variety in use. The Marsha concept, short for Mars habitat, envisages a 3D-printed biopolymer vertical home designed for life on the red planet. Pisces, a LED-light emitting device, that attracts the size and species that fishermen are licensed to catch while repelling non-targeted species, is enabled by plastics as are the X-box Adapative Controller, redesigned making no assumptions about players' capabilities, and Thumy, part medical tool and part toy, that helps children with diabetes inject themselves. But perhaps the most poignant winner in the context of this book is The Ocean Cleanup Array. At the time (2013), the designer was only 19. He uses plastics structures to clean up, collect and dispose of the plastics of the Great Pacific Garbage Patch. Trials and development are ongoing, but I am encouraged by these designs exploring the potential of the 'problematic' material to contribute to the solution. We have no alternative but to make plastics work for sustainability.

And it is coming.

Increasingly, design companies are opening up with a focus on recycled products. I have just learned of SMALLRevolution. A Danish company cleverly occupying one of the most challenging spaces in a sustainable plastics economy—the design space. SMALL designs desirable furniture from the recycled plastics that right now are piling up in re-cycling plants throughout Europe. Its products are fully recycled and recyclable because they do not use glues. They also select bright colours and by so doing make desirable products as opposed to the many brownish/grey recycled plastics products we are used to by now. By doing so, they close the gap between re-cycling plants and consumers.

Today, I write from our summer house in which we are in COVID-19 hiding. Arriving here I was shocked to find that plastics are not sorted in this part of Denmark. Sorting plastics has in the five years since the conference become a daily routine in Copenhagen where we live.

This book presents a variety of views, both academic and practical, and does so in a balanced manner describing both positive and negative influences of plastics in our lives. It makes a timely contribution that, read with open minds, will contribute to creative approaches to the 'plastics problem'.

Copenhagen, Denmark   Kigge Mai Hvid

ACKNOWLEDGEMENTS

Many people have contributed to this volume: its incubation, its development and its fruition as something you can hold in your hand.

First, my thanks go to the Arts University of Bournemouth (AUB) for its support not only during the development of the book but also for accepting my proposal in 2006 that its existing accredited 'Design Museum Collection' should build on its strengths and focus on design in plastics. I would like to thank especially the then Deputy Vice-Chancellor, Professor John Last, who commissioned the review and as a result gave me a new lease of life as well as setting the transformation of the collection in train. Consequently, the Museum of Design in Plastics (MoDiP) was established in 2007. This has been a turbulent time for many museums. University museums in particular have experienced cutbacks but conversely MoDiP has thrived and to that end we have been greatly assisted by the University's senior management, especially Professor Stuart Bartholomew and Professor Emma Hunt, respectively vice-chancellor and deputy vice-chancellor for much of the period. Their belief in the concept and the support they have provided has been a source of constant encouragement and strength. I would also like to thank our current vice-chancellor, Professor Paul Gough, for the commitment he has already shown to MoDiP.

Then, I would like to thank all those who took part in the 2015 conference, *Provocative Plastics: Design in Plastics from the Practical to the Philosophical* on which the book is based. They have become stalwart collaborators. I thank them for participating in the conference in the first place, sticking with it and providing their extended chapters. I am most grateful for their inspiration, dedication, and co-operation.

Accolades are due also to my wonderful MoDiP colleagues. We are a small, dedicated team and I have been very fortunate to work with such a motivating and caring group. I thank in particular Pam Langdown, who nurtured the collection single-handedly for many years and shares her knowledge with such generosity; the Curator Louise Dennis and the Collections Manager Katherine Pell, both of whom have kindly sorted me out on those many occasions when I have needed help; and Julia Pulman for her empathetic presence.

Valerie Lodge, the AUB's research manager, has also provided support in so many ways over the years. I am especially grateful for her tactful but spurring-on queries concerning the progress of this publication. They have been key to its realisation.

My friend, Professor Marcia Pointon, has also generously provided crucial advice at many stages before and along the volume's gestation. I thank her particularly for her encouragement and for coming to the rescue at vital moments.

These acknowledgements would not be complete without thanking my publishers for sharing their wisdom and knowledge in guiding me through the sometimes arduous process of publication. I owe a particular debt to Lina Aboujieb for her initial enthusiasm and ongoing open mind and to Julia Brockley and Emily Wood, who answered my many questions promptly and with kindness and helped in so very many ways. Their contribution has been invaluable.

Lastly, a word of thanks to my husband, James Rees, who has developed considerable skills in sub-editing and in editor calming.

Susan Lambert

CONTENTS

Notes on Contributors

Eric Bischof has more than 25 years of experience in the chemical industry, including positions in manufacturing, product management, supply chain management and business process redesign. He is the inventor of the energy efficiency methodology licensed under the trade name STRUCTese. Eric headed the Sustainability initiative, Chemie[3], of the German chemical association (VCI). Today, he serves as Vice President Corporate Sustainability at Covestro.

Deborah Cane ACR, FIIC was formerly collections care manager at Birmingham Museum Trust. Her responsibilities included managing the conservation and collections management teams and the Museum Collection Centre as well as the city public art portfolio. In 2016, she became conservation manager for Sculpture and Installation Art at Tate where she oversees work programmes across Tate's five sites.

Sebastian Conran has a long career as a leading British designer and is managing director of Sebastian Conran Associates (2009–). He is a founding trustee of the Design Museum and has strong links with higher-education as Designer in Residence at University of Sheffield, visiting professor at Bristol Robotics Lab, past visiting professor at Central St. Martins and lecturer at the Royal College of Art. He has received many academic honours, industry awards and patents for his design and innovation work.

Russell Gagg has held chartered status with the Royal Institute of British Architects since 1998. He has worked in practice across the United Kingdom, Asia and Australasia including at Victoria University of Wellington, New Zealand, where he was lecturer and studio leader on the BA (Hons) and MA Architecture programmes. He joined the Arts University Bournemouth in 2006 where he is course leader of Interior Architecture and Design.

Maria Georgaki is a design historian and teacher. She is particularly interested in the history of design education and the educational affordances of material culture. Using the Camberwell ILEA Collection as a case study, her doctorate research, undertaken at the Arts University London, examined how handling collections foster learning through objects. She works as a primary school teacher.

Kirsten Hardie is a UK National Teaching Fellow and has an international reputation for innovative pedagogy. She is an associate professor at the Arts University Bournemouth where she teaches Graphic Design History and Theory/Contextual Studies. Her professional activities include external examinerships, peer review, curatorship, consultancy and journal editorial work. She is co-president, International Federation of National Teaching Fellows. Her research includes object-based learning, flock, packaging, graphics, and artificial flowers.

Richard Hooper is an associate professor in the Fine and Applied Art Department at Liverpool Hope University. He has a digital studio and sculpture practice and exhibits work in the United Kingdom and abroad. His work is in the permanent collection of the Los Angeles County Museum of Art.

Brenda Keneghan has a primary degree in Chemistry from University College Cork and a PhD in Materials Science from Queen Mary University of London. She worked for several years in academic research in polymer chemistry before joining the Conservation Department of the Victoria and Albert Museum in 1993. She is the editor of *Plastiquarian*, the journal of the Plastics Historical Society.

Susan Lambert worked for many years at the Victoria and Albert Museum, ultimately as Keeper of the Word and Image Department. Highlights included organising the V&A's first global 20th Century Gallery and setting up its Contemporary Programmes. In 2007, she was instrumental in developing an existing artefacts collection at the Arts

University Bournemouth into the Museum of Design in Plastics (MoDiP), where she is professor and chief curator.

Joanne Lee is an artist/writer/publisher whose research explores everyday things and places. Her creative and critical work is often developed through the Pam Flett Press, an independent essay serial. She is a senior lecturer in the Department of Art and Design, Sheffield Hallam University.

Gerson Lessa is a PHD professor in the undergraduate program in Industrial Design in the Universidade Federal do Rio de Janeiro, Brazil. He is focused in Design History and Material Culture Studies. As an avid collector of design in plastic materials, all his work is based on a hands-on approach.

Stefan Lie is Senior Lecturer in Product Design at the University of Technology Sydney. He conducts research in material interaction with the focus of developing novel ways for people to interact with physical products. Lie specifically investigates new ways of developing form, the social meaning and value of materials to people and our environment and the creation of more sustainable materials for product design practice.

Flora McLean is head designer of the fashion label House of Flora, which she set up in 1996, a year after graduating from the RCA. The label specialises in avant-garde headwear for the catwalk, fashion campaigns and collectors as well as a commercial range of statement fashion jewellery and eyewear pieces. She is also Senior Tutor in Footwear, Accessories, Millinery and Eyewear at the Royal College of Art.

Susan Mossman has specialised in the history and preservation of plastics, with additional research interests in materials science, archaeometallurgy and museological studies. Her professional career had been spent at the Science Museum in London UK, as a curator, project leader and now as an honorary research associate.

Berto Pandolfo is an industrial designer with over thirty years' experience. As an academic design practitioner, Berto contributes through collaborative, industry-focused research and teaching. His key expertise is in traditional and emerging technological and material applications, which positions him at the forefront of current design research practice. He is Senior Lecturer in Product Design at the University of Technology Sydney.

Tone Rasch is a curator at the Norwegian Museum of Science and Technology in Oslo. She is an art historian, with special interests in textile industry, fashion, and plastics, and has for years worked with the business archive from the Norwegian textile mill Hjula, preserved at the museum.

Mark Suggitt began his career as a social history curator at Salford Museum and Art Gallery. Subsequently, he worked at a number of museums including as director of St Albans Museums, head of Bradford Museums and Galleries and director of Derwent Valley Mills World Heritage Site. He was a board member of the Museums Association, the International Council of Museums, World Heritage UK and chaired the Social History Curators Group and the Plastics Subject Specialist Network.

Roderick Walden undertakes research into the contemporary methods, practices, and systemic conditions of professional industrial design. His experience as an industrial designer for local and international manufacturers informs his academic focus, to develop design methodologies and theoretical frameworks for knowledge-intensive industrial design that situates prototyping as a central research device. Roderick is Lecturer in Product Design at the University of Technology Sydney.

List of Figures

CHAPTER 1

Introduction

Susan Lambert

Plastics are an especially important group of materials because of their fundamental role in shaping our world. There are literally thousands of different plastics, each with their own composition and characteristics. For this reason in this publication we use the word plastics in the plural in order to indicate the plurality of the material but with three exceptions: plastic bags, plastic bottles and plastic flowers, on the basis that they are everyday phrases established in our consciousness in the singular.

The plastics materials group includes many of the cheapest but also some of the most highly engineered and complex materials available. For nearly fifty years they have been the most used materials group (Cascini and Rissone 2004). Nonetheless plastics are controversial and provoke strong opinions based on an uneasy mix of emotion, experience and knowledge. For some they are beautiful, adaptive and creative; for others they are unpleasant, inauthentic and destructive. The purpose of this book is to present a range of evidence-based discourses on attitudes to plastics from the positive to the negative and thus provide a counterpoint to much of the recent literature. The intention is to contribute a more balanced appraisal of the value of plastics than has been achieved hitherto.

S. Lambert (✉)
Museum of Design in Plastics, Arts University Bournemouth, Poole, UK
e-mail: slambert@aub.ac.uk

The word 'value' is almost as ubiquitous as plastics and much has been written about theories of value. As Daniel Miller has pointed out, it is a word with an extraordinary semantic range in the English language: 'On the one hand it can mean the work involved in giving a monetary worth to an object ... and thereby becomes almost synonymous with price. On the other hand it can mean that which has significance to us precisely because the one thing it can never be reduced to, is monetary evaluation, for example the value we hold dear in relation to family, religion and other inalienable possessions' (Miller n.d.).

An aspect of this value conundrum that preoccupied the eighteenth-century Scottish economist, philosopher and author, Adam Smith, is the different values attached to water and diamonds. We put a high value on diamonds, which are inessential to human life and a low value on water, without which humans would die (Smith 2009, i:iv). This compares with the cheapness of many plastics and, arguably, how they are valued. Yet the simple plastics bucket, introduced in the 1960s, has been the agent of cleaner water that has dramatically improved the life expectancy of the 46% of the world's population who still do not have access to piped water at home (World Health Organisation n.d.).

Contributors to this publication are drawn from a range of disciplines and thus it encompasses concepts of value influenced by different bodies of theory. However, at the heart of our discourse is the theory of utility, which has a long evolving history stretching back to Aristotle but was clearly propounded independently during the 1870s by the economists, William Stanley Jevons, Carl Menger and Léon Walrus. It argues that value arises from the relationship things have with users' needs, in this context whether as makers or as consumers, and is not inherent in the things themselves. Therefore, with changes in the relationship between user and object, value can rise or disappear. It has no independent existence but is entirely dependent on a judgement made by a particular person or group of persons in a particular context (Milward 2000). In this respect, Bourdieu's writings on cultural capital, and the ascendancy in matters of taste of those who have cultural capital, is significant particularly in appreciating the intricacies surrounding such judgements. To what extent are they influenced by the extent of the particular judge's accrued breadth of cultural capital (Bourdieu 2010, 2)?

CONTEXTUAL REVIEW

This volume is based on papers given at the conference, *Provocative Plastics: Design in Plastics from the Practical to the Philosophical,* held at the Arts University Bournemouth in September 2015. Accounts that explore this broadly cultural contribution of plastics, rather than the technical or conservation issues they raise (Keneghan and Egan 2009; Koestler 2017; Quye and Williamson 1999; Shashoua 2008) are comparatively few and far between. Moreover, often their purpose has been either to celebrate a plastics material's capabilities or to suggest ways of making plastics more effective. An early and remarkably extensive homage to plastics is provided by the American Craft Council's 1968 exhibition which was divided into 'Plastics craft', 'Plastics as art material' and 'Plastics industry and trade' (American Crafts Council 1968) and the same year the Flint Institute of Arts organised an exhibition *Made of Plastic* featuring the work of leading contemporary artists such as Andy Warhol, Claes Oldenburg, and Roy Lichtenstein (Hodge 1968). These were followed in 1969 by an exhibition *A Plastic Presence,* which explored plastics as a material that has extended the artists' material range through sample works by forty-nine sculptors living in the United States and Canada (Atkinson 1969). Sylvia Katz's ground-breaking work *Plastics: Design and Materials,* published in 1978, is more analytical and fulfils its title brilliantly through an account of the evolution of form in plastics design (Katz 1978). Another notable predecessor is RAPRA's Technology's conference, *The Art of Plastics Design,* held in Berlin in 2005 at which leading British designer Sebastian Conran (a contributor to the current volume) in his keynote speech played on the semantic range of the term 'value' presenting the role of the designer as to 'create value … our principal task is to maximize the primary value of "desire", and to thereby positively influence the price-driven one …' (Conran 2005, 2). However, significant as this conference was, again its ambition was different from that of *Provocative Plastics* and the current publication, being predicated on plastics being, if not necessarily a good thing, at least a given. Hence papers were focused on specific approaches to designing with plastics, the development of plastics brands and ploys to enhance the attractiveness of plastics products.

Arguably it is the equivocal nature of plastics that makes them provocative. Possibly the first writer to signal this equivocality was the French cultural critic and moral philosopher, Roland Barthes, in his much quoted in this volume and elsewhere seminal but brief appraisal of an unnamed plastics exhibition in 1957, where he experienced manufacture in plastics from the raw material to the product. On the one hand he found plastics 'in essence the stuff of alchemy' and 'a miracle substance'. On the other he found them 'disgraced', 'imitation' and 'prosaic' materials, their saving grace being that 'for the first time, artifice aims at something common, not rare' (Barthes 2000, 97–98). The ambivalent contribution of plastics featured also in a short article of 1984 entitled 'The Plastics Man' by the American historian Robert Friedel. In this he described Leo Baekeland, the inventor of the first fully synthetic plastic, as 'the father to the family of versatile, exasperating, indispensable materials' that had made 'the very feel of the twentieth century unique' (Friedel 1984, 49). The excitement the Italian designer Ezio Manzini felt towards plastics is expressed in the title of his 1989 book, the *Material of Invention*; however, he has also presented them in a more questioning light. For example, in a paper of 1992, he stated: 'Plastics, due to their having been the privileged medium of quantitative development, have obviously been profoundly involved in the problems that this "discovery of the limits" has imposed on our common attention. The remarkable combination of formability and cheapness that typifies these materials, and that led them to become a decisive factor in the multiplication of products, and in the reduction of their life cycles, has implicated them in the issues of energy and material waste within the current system of manufacturing and consumption' (Manzini 1992).

A significant text that underpins most cultural studies in plastics, including the current volume, is Jeffrey Meikle's illuminating and balanced *American Plastic*. However, as the title makes clear, his canvas is limited to the United States and it was written over twenty years ago. At that time, the magnitude of the global threat plastics posed was less evident and 3D printing, which is transforming who uses plastics, how and why, had scarcely been invented (Meikle 1995). Other more journalistic but nonetheless valuable cultural histories of plastics, also from the American perspective, are Stephen Fenichell's *Plastic: The Making of a Synthetic Century*, 1996, and Susan Freinkel's *Plastic: A Toxic Love Story*, 2011. However, their concerns with the environmental consequences of plastics have encouraged an engaging but somewhat negative view. Fenichell describes plastics as 'a white trash connoisseur's enduring delight. Think Ban-lon, Orlon, Corfam, Lycra. Leatherette' (Fenichell 1996, 6–7). Freinkel chillingly describes our feelings towards plastics as a complicated mix of dependence and distrust 'akin to what an addict feels towards his or her substance of choice' (Freinkel 2011, 8).

Additionally, there are two important essay collections edited by Sparke (1990), and by Gabrys et al. (2013) that complement the current volume and, like it, benefit from a variety of viewpoints. Sparke, a pioneer writer in this subject area, maps the intellectual territory as it was in 1990 through a useful collection of key texts by thinkers and plastics experts, past and present. Gabrys et al. provide a more recent, ground-breaking, interdisciplinary discourse on the material politics of plastics. It elucidates the complexity of this near infinite materials group through a processual approach. In particular, Hawkins reminds us that the value attributed to plastics is not intrinsic to the material but rather 'enacted': 'the economic capacities of plastic emerge in specific arrangements and processes, in which the material interacts with any number of other devices—human and non-human—to become valuable' (Hawkins 2013, 49).

As well as texts that address plastics specifically there is a large underpinning and overlapping literature on materiality, sustainability and design, all of which impact on analysis of the value of plastics. As a sample, Bruno Latour developed the concept of the actant, a source of action that can be non-human as well as human, in his exploration of the philosophy of environmental conflicts (Latour 2004, 237). Plastics are likely to have been among the actants he had in mind. Jane Bennett took up the term in *Vibrant Matter* in which she explores the agency of non-human resources reminding us that trash in landfills is not ' "away"… but generating lively streams of chemicals and volatile winds of methane' and asks 'would patterns of consumption change if we faced not litter, rubbish, trash, or "the recycling", but an accumulating pile of lively and potentially dangerous matter?' (Bennett 2010, vii–viii). *The Culture of Nature in the History of Design* explores the dialogue between design and nature presenting it, in tune with plastics in this volume, as both part of the environmental problem and part of the solution (Fallan 2019). *Cradle to Cradle. Remaking the Way We Make Things (Patterns of the Planet)* written by a German chemist and a US architect posits a new pattern of making in which the valued concept 'reduce, reuse, recycle' is replaced by the breaking down of the obsolete into either environmentally friendly biological 'nutrients' or technical 'nutrients' retained within closed loop industrial cycles (Braungart and McDonough 2002). It presents a future for which we must strive.

Objectives

The purpose of this book is distinct from and complementary to this literature. It makes no attempt to provide another history or overview of plastics' politicised place in our globalised world. Nor, in spite of our wish that it should be considered timely, does it have a particular focus on contemporary agendas such as climate change, ocean pollution or responsible design. Nonetheless, these subjects have immense significance in terms of its focus and thus inevitably enter into the discussion from a variety of viewpoints. However, the book's objective is rather to provide a balanced exploration of how plastics have been valued over time, specifically from two human perspectives: as a medium for making and in societal use. The first explores the multivalent nature of plastics' materiality and their impact on creativity through the professional practice of artists, designers and manufacturers. The second explores attitudes to plastics and the different values that can be applied to them through the current research of design, materials and socio-cultural historians. This introduction locates these contributions within each author's frame of reference and in relation to a wider narrative of how plastics have changed our world for better and for worse.

The book is unusual in the range of its tone arising from the variety of experience of its authors. Some have hands-on knowledge of working with plastics. As a result, their perceptions stem from anecdotal experience that, nonetheless, because of their practical knowledge of working with plastics as a means of livelihood, contribute meaningful testimony to the wider picture of plastics' value. Others have researched their subject from specific theoretical standpoints and provide more traditional academic arguments. Some themes are common across chapters. Indeed there is a synergy across the texts in the two sections but from different perspectives. It is the interdisciplinary approach that sets this compendium apart. It brings together a variety of voices to unpick values attached across time to this paradoxical materials group, as their unique properties lead them to play an ever more essential part in our lives whilst simultaneously their ubiquity creates an ever-greater problem that we must solve.

PLASTICS IN PROFESSIONAL PRACTICE

The early inventors of plastics valued them just because they could substitute for natural materials. Alexander Parkes's Parkesine, an early form of cellulose nitrate, is often cited as the first semi-synthetic plastic. In Parkes's announcement of his invention at London's 1862 International Exhibition, he made it clear that he was proud that it was capable of 'the most perfect imitation of Tortoise-shell, Woods, and an endless variety of effects' (Plastics Historical Society n.d.). John Wesley Hyatt's later development of celluloid,[1] patented in 1869, is said to have been motivated by a reward of $10,000 for the discovery of a material that could take the place of ivory in the manufacture of billiard balls (Meikle 1995, 10). The British Xylonite Company, manufacturers of cellulose nitrate, founded in 1877, were so proud of the imitative capacity of the material that it marketed its material under the trade name 'Ivoride' with a tortoise and an elephant as its trademark. A sample of Ivoride bearing the logo can be seen in the Plastics Historical Society's collection (Museum of Design in Plastics n.d.).

Hyatt, moreover, went as far as to make the case that his celluloid combs improved on those made from natural materials. In an 1877 patent application, he claimed: 'Heretofore, as far as known, combs have been made of wood, metal, horn, hard rubber, and an imperfect and unsuitable variety of pyroxyline.[2] Such combs possess the defects that are incident to the material of which they are made when employed for such structures.' He goes on to catalogue their various deficiencies declaring, 'The object of this invention is to furnish a comb that possesses more valuable qualities than any heretofore constructed, and which at the same time is free from all the defects' (Google Patents n.d.).

Leo Baekeland also saw Bakelite, the trade name he gave to his invention of the first fully synthetic plastic, phenol formaldehyde, as an improvement on nature. Shortly after its discovery in 1907, he described his process as 'a laboratory reaction, which seems to run along lines parallel to those of the delicate physiological changes which occur in plant life' but the Bakelite reaction is 'quicker and less expensive' (Meikle 1995, 48). The scientific commentator Edwin Slosson, writing in 1919, defined creative chemistry as 'the art of substances not made in nature'. He saw in synthetic plastics the building blocks of utopia: 'Gradually then he will substitute for the natural world an artificial world, molded nearer to his heart's desire. Man the Artifex will ultimately master Nature and reign supreme over his own creation ... It is by means of applied science that the earth can be made habitable and a decent human life made possible' (Slosson 1919).

It could be seen as the misfortune of plastics, such a wondrous family of materials in the eyes of some, to have been developing at just the moment the truth-to-materials doctrine was being promulgated first by the Arts and Crafts movement and then by modernism. This doctrine was itself a response to the industrial developments of which the plastics industry was a part. Its essence was that materials should be used in ways determined by the character of the materials themselves. There was also a commitment to craft working and the honest display of construction methods. Ruskin named among his three 'Architectural Deceits': 'The suggestion of a mode of structure or support, other than the true one ... The painting of surfaces to represent some other material than that of which they actually consist (as in the marbling of wood)... [and] The use of cast or machine-made ornaments of any kind' (Ruskin 1849, 39). These ideas were further developed by William Morris, whose hierarchy of materials based on the criterion of 'nobility', presented stone as the most 'noble', followed by wood and then brick, which he described as a 'makeshift' material (Morris 1892). Was his problem with bricks that they are manmade? Plastics are another step away from this concept of 'nobility' because not only are they manmade but also predominantly made from substances created in a laboratory. These attitudes lingered well into the twentieth century with the sculptor Henry Moore, stating, 'Each material has its own individual qualities ... Stone, for example, is hard and concentrated and should not be falsified to look like soft flesh ... It should keep its hard tense stoniness' (Moore 1934) although subsequently he relented, 'I still think it important but it should not be the criterion of the value of a work—otherwise a snowman made by a child would have to be praised at the expense of a Rodin or a Bernini' (Moore 1970).

The concept of truth-to-materials is fundamentally at odds with plastics, their defining nature being that they have no specific indigenous character to which to be true. The Science Museum's pioneering exhibition *Plastics Materials Industry* of 1933 proudly traced 'the course of the plastics industry from the various raw materials used in the production of plastics to the final product of artificial wood, china, horn and metals'

(Plastic Materials 1933). The exhibition of *Plastics* that opened at the Institute of Modern Art, Boston, late in 1940, described plastics as being 'at once wood, rubber, metal, glass and yet none of them.' The ramification drawn in this instance was that 'the designer cannot depend on the inherent nature of his material for inspiration but is obliged to draw from the heritage of the "pure" materials which plastic resembles' (Meyer 2013, 241–243). Barthes, however, saw a different future for plastics and it is his concept of plastics as 'the very idea of its infinite transformation' (Barthes 2000, 97) that has made plastics much more at home in today's postmodern world which challenges established definitions of 'good' art and design and embraces variance.

Plastics materials proved inspirational to designers even before they could deliver their promise. This was the case with Saarinen's famous tulip chair, one of the first one-legged chairs to be produced, which began manufacture in 1956, the year before Barthes's comments. Saarinen designed its fluid shape in clay and wanted the chair to be moulded in a single plastic. However, to make it structurally sound, the glass-reinforced plastics seat is supported on an aluminium pedestal coated in a polyamide to give the appearance of a single material. Thus, Saarinen forced his unified vision onto these new materials, commenting, 'All the great furniture of the past from Tutankhamen's chair to Thomas Chippendale's have always been a structural whole ... I look forward to the day when the plastic industry will advance to the point where the tulip chair will be one material, as it was originally designed' (V&A n.d.). It is the ability of plastics to inspire a new approach to artistic practice that is explored by Russell Gagg in this volume and leads to his conclusion that 'plastics are the new magic'.

The centrepiece of the 1968 American Crafts Council *PLASTIC as Plastic* exhibition was a sculpture in polymethyl methacrylate by Alexander Calder. It had been the winner out of 250 entries in a competition set up by Rohm and Haas, the manufacturers of this material under the name of Plexiglas, as part of their promotion at the 1939 New York World's Fair. The jurors of the competition, which included James Johnson Sweeney, director of Painting and Sculpture at the Museum of Modern Art, New York, praised the work for its exploitation of Plexiglas's ability 'to carry light around curves' and 'from a source to an outlet without giving off light in transit' making the sculpture 'edge-lit' and thus making 'light an organic part of the design' (Beinecke Rare Book and Manuscript Library n.d.).

In fact, a number of modernist artists had previously used plastics as a creative medium in their own right. Naum Gabo, who taught at that bastion of modernism, the Bauhaus in Germany, in the late 1920s, stands out as an early example. He made his first plastics sculpture in 1920, working first in cellulose nitrate and then, as new materials became available, in cellulose acetate followed by polymethyl methacrylate, better known by its trade names, Plexiglas and Perspex, and also known as acrylic. The fate of one of these sculptures is discussed in Chapter 13. Gabo was drawn to plastics specifically because, as a construct of the chemical industry, they acted as a device to divorce his art from what had gone before. Additionally, although in appearance similar to glass, they did not break. Their inherent plasticity made them the perfect material for him to cut and bend into his vision of a conduit to 'a more perfected social and spiritual life conducted and based upon universal principles' (Hammer and Lodder 2000, 260 note 239).

The sculptor Richard Hooper contextualises his own practice through a study of plastics' contribution to sculptural vocabulary thereby demonstrating the unusual variety of effect and meaning achievable with this materials group. His focus however is, like Gabo's was, on plastics' translucency. His work investigates the potential of luminous acrylics worked with computer-aided design and computer-aided manufacturing (CAD/CAM) methodologies to play a similar role in our culture as was played by stained glass windows in medieval times: that is, the representation of spiritual *transcendence*.

The introduction of semi-synthetic fibres mirrored that of solid plastics. The earliest commercially successful example, Chardonnet silk, made public at the Paris International Exhibition of 1889, was marketed both as an artificial imitation and as cheaper than the natural material. However, it proved to be outstandingly flammable (Painter and Coleman 2009, 10). Its more stable successor in the market place, known variously as viscose or rayon depending on the country, further discussed in Chapter 11, was also troubled. Although cheaper than natural silk, it did not possess the same elasticity. This led, as hemlines rose, to wrinkled knees when worn as hosiery and thus a questionable reputation. For this reason, it was important for the American Company, DuPont, when announcing the first fully synthetic fibre to which they gave the trade-name, nylon, to disassociate it from the existing semi-synthetic fibres. They succeeded by replacing their normal scientific approach to marketing with a more domestic and feminine one. Young women assuming various roles were 'exhibited' actually wearing nylon stockings. As Meikle put it, 'Giving lip service to an agenda of chemical utopianism, a male lecturer introduced her by observing she was dressed "by chemistry from head to toe," from her Lucite jewelry and rayon dress to the cellophane bows and Pyralin heels of her shoes—and, of course, her nylon stockings' (Meikle 1995, 144). Among the first actually to experience wearing 'nylons' were staff at DuPont in early 1939. They reported the effect they had on how they felt as well as how they looked. One said: 'My legs felt different ... I felt like kicking them up in the air ... somehow, they made you feel you had gone to town on yourself, like spending fifty bucks on a hairdo, a facial, a bottle of perfume ... when all you'd done was put on a new pair of stockings' (Mosley 1980, 369). Since then nylon and later synthetic materials such as Lycra have continued to mould not only women's bodies but also contributed to the construction of their identities. This is addressed in this volume by the feminist milliner Flora McLean, for whom plastics is a material of rebellion. She analyses both how she and others use plastics, thereby making the case for plastics being a material that has not only the ability to transform women physically but also to enable them to find their own individual routes to power and fulfillment.

Plastics only became a regular part of everyday domestic life after World War II. However, during the war years, the industry grew at a great pace. Once peace was restored, there was a need to find a new outlet for the range of plastics materials that had been developed. In October 1947, the American magazine *House Beautiful* devoted some fifty pages to 'Plastics ... a way to a better more carefree life'. It explained, 'The greatest virtue of plastics is that they *are* different, so it is possible to select from the many the one plastic that will do a given job best ... No plastic is any better than the use to which it is put ... whenever a plastic performs imperfectly, you may be pretty sure that it's because someone—the manufacturer or the consumer—has tried to make it do a job it was never supposed to do' (Gough 1947, 121). The frequency with which this occurred in the 1950s and early 1960s has been a significant contributor to a widespread sense that plastics are cheap and nasty materials. Even a designer such as Sebastian Conran, of significance not only because of the quantity and quality of the products he has designed but also because his career has spanned the period since plastics became the most used material in design, would like to 'de-plastify' the world. His chapter, drawing on his experience as a designer for different companies including leading Sebastian Conran Associates, provides a telling insight to the reasons why designers work with plastics and the quality/cost/value dilemmas they encounter. He explains how his design team sets out to mitigate plastics' aesthetic impact as 'plastic' by the propriety of their use and their teaming with other complementing materials that have been traditionally valued more highly concluding that 'deplastification is an important consideration in achieving a product that will be cherished'.

Plastics are worked in a wide variety of ways including with hand-held tools to the highly mechanised process of injection-moulding, globally the manufacturing technique most widely used since World War II. It has transformed manufacturing from an intrinsically craft process to one of almost unlimited mass production. However, the tooling required is expensive. As a result, it is only cost efficient across a very large number of units, contributing to the large production numbers associated with plastics. 3D printing, however, has the potential to turn this status quo on its head. The chapter by Stefan Lie, Berto Pandolfo and Roderick Walden explores the high quality and glossy finishes for which plastics have been valued and the reasons why 3D printing, a relatively new fabrication process, which still struggles in this respect, is, nonetheless, challenging the existing manufacturing paradigms as a highly valued means of production.

One of the most provocative and equivocal mass-produced plastics products is the plastic carrier bag. Its invention as we know it today, with integral handles, is usually attributed to Sten Gustav Thulin who was granted a patent for a 'bag with handle of weldable plastic material' in 1965 (Espacenet Patent Search n.d.). It is now estimated that some 5 trillion plastic bags are used a year and a wide range of pressure groups have fought to have them banned with some measure of success. Bangladesh was the first country to ban them in 2002. The same year Ireland introduced a charge of 0.15 Euro (now raised to 0.22 Euros); South Africa banned them in 2003 and San Francisco became the first US city to ban them in 2007. Wales introduced a 5p charge in 2011 with England and Scotland following suit in 2015 (Big Fat Bag n.d.). The justification for these bans and levies seems to have had less to do with issues relating to sustainability and more to do with the plastic bags ability to litter the landscape. This had been a particular problem in Bangladesh, where drainage systems clogged with plastic bags were a major contributor to flooding in the monsoon season (Onyanga-Omara 2013). However in Ireland it was related at least to a certain extent to the tourist industry. The Irish Environment Minister Noel Dempsey said on introduction of the charge: 'plastic bags are a visible and persistent component of litter pollution in towns, throughout the countryside and along the coastline … Their prevalence suggests that we are careless of our environment, and of the environmental image we convey' (INCPEN 2008). Mark Murray, executive director of Californians against Waste, said: 'They're so aerodynamic that even when they're properly disposed of in a trash can they can still blow away and become litter' (Mieskowski 2007). Indeed, before their ban in South Africa, they were known as the national flower because of the number that used to fly into bushes and trees (BBC News Channel 2003). In sympathy with the concept of the bag as a flower, Joanne Lee's chapter asks us to look again at this capacity of the plastic bag and consider both its artistic contribution to our highways and byways and its potential as 'a creative and critical metaphor for the process of [her] research and its articulation'.

If your concerns are energy consumption or climate change, it can be argued that plastic bags are more eco-friendly than their more traditional alternatives. Their manufacture requires little energy compared with that of paper bags. They are also made from a waste product of the petrol industry and do not require the cutting down of forests, vital creators of oxygen. Marks and Spencer reported that it took three times more lorries, and therefore three times as much petrol, to deliver the same number of paper as plastic bags to their Irish stores (Scottish Plastics and Rubber Association 2013). In 2011, the UK Environmental Agency published a lifecycle assessment of a range of bags and concluded that thin high-density polyethylene bags cause less damage to the environment than any other kind of bag (Edwards and Fry 2011). Eric Bischof, speaking from the perspective of a manufacturer who is aware of the importance of the bottom line, discusses both the energy intensive aspects of plastics and their capability to reduce resource demand and environmental impact. His chapter draws especially on the contribution of innovative start-up companies focused on conserving our planet. He posits that an excessive focus on one lifecycle may be counter-productive and that to assess the true impact of plastics on sustainability requires a holistic approach driven by 'the three Ps: planet, people and profit'.

Plastics in Societal Use

Plastics provoke passionate responses. The Americans, artist Andy Warhol, and novelist Norman Mailer, are famous in this respect. Warhol's statement made in the 1960s: 'I love Los Angeles. I love Hollywood. They're beautiful. Everybody's plastic, but I love plastic. I want to be plastic' (Thomas 2005, 13) was intended to be controversial. His heightened veneration for plastics is matched by Mailer's hyperbolic virulence towards them. In 1983, Mailer called plastics 'the excrement of oil', attributed violence in society to the malign influence of plastics on the young, and described plastics as 'the social equivalent of cancer' (Mailer 1988, 321). Recently, an informal focus group held with my extended family elicited plastics 'stink'. The only person to speak up in their favour was the oldest member who pointed out he owed his sight (spectacles and artificial lenses post-cataract operations), hearing (hearing aids) and heartbeat (pacemaker) to plastics of one kind or another. Others might have mentioned hip replacements and key-hole surgery.[3]

However, even if we do not rely physically on plastics, most of us clean our teeth with plastics toothbrushes, wear clothes next to our skin made from mixed fibres containing synthetics, and drink from plastic bottles; we do not give a second thought about the essential role plastics play in the make-up of domestic, medical, transport and electrical and digital products. Indeed e-waste, unwanted electrical and electronic equipment is currently, at 30 million tonnes annually, the world's fastest-growing type of waste, half of which is made up of discarded head phones (Museum of Design in Plastics 2019). According to ERI, specialists in electronic recycling, Americans spent $71 billion on communication equipment in 2017, nearly five times that which was spent in 2010 (Semuels 2019). Given our dependence on plastics, it is curious that they tend to provoke such extreme responses, indeed more extreme than is the case with other petrochemicals. Is it, as implied by Mailer, because their role in our lives is more intimate getting 'into every single pore of productive life' (Mailer 1988, 321)?

Whatever we think of plastics, they are now an intrinsic part of our lives in most parts of the world, but surprisingly that was not the case when plastics began to feature in popular culture. For example, the cartoon character, Plastic Man, a crook who while on the run became exposed to a chemical which turned him into plastics, appeared in 1941 (Cole 1941) when plastics were more to do with the war effort than domesticity. Three years later, Walt Disney's 'Plastics Inventor', in the persona of Donald Duck, took instructions from the radio's 'plastic hour' and baked a plastics plane from household junk (Disney 1944). Thus, already plastics were associated with trash. In May 1943 the US magazine *Life* subtitled an article on plastics, illustrated by a photograph of a worker putting the finishing touches to a clear plastics bomber nose, 'War makes gimcrack industry into sober producer of prime materials' (Anonymous 1943). Perhaps plastics' reputation suffered from the fact that they were made by the same chemical companies, such as DuPont, that made bombs (Ndiaye 2007). It was as if the idea of plastics were greater than their reality yet there was also already a sense they would not live up to expectations. Plastic Man proved his plasticity by, on the one hand, being a 'reformed' character that acted on behalf of the FBI and, on the other, keeping his 'tacky' contacts in the underworld (Thompson 2008). Donald Duck's plane defaulted to dough in the rain, re-forming into a parachute as it plunged ground wards, reverting to piecrust on landing. In the first chapter of this section, social historian Mark Suggitt explores such cultural references since World War II looking particularly at how plastics have been popularly presented and the part they have played in shaping and enabling popular culture.

The general public first became aware of this new kind of material (now grown into the large materials group known as plastics) when they found, towards the end of the nineteenth century, that what looked like luxury goods were becoming increasingly within their financial reach. To quote Meikle: 'By replacing materials that were hard to find or expensive to process, celluloid democratized a host of goods for a consumption-orientated middleclass' (Meikle 1995, 14). This first plastic was, as Robert Friedel reflected, 'treated not as a cheap utilitarian stuff but rather as a raw material for artistry and ornament' (Friedel 1987). The availability of these celluloid combs is said to have contributed to the fashion for long hair and by the turn of the century contemporary hairstyles often required 'a couple of pounds of celluloid' to hold the hair in place. Moreover, it was possible to manufacture them in the much greater numbers required to meet the demand to which their democratisation gave rise. As a result, when bobbed hair came into fashion in the early 1920s, the celluloid industry was seriously challenged to find other uses for the material (Dubois 1972, 47).

Celluloid, or cellulose nitrate as it is known in the United Kingdom, was able to create such convincing look-alike products that even after the passing of a century it can be hard to tell whether they are real or imitation, as discussed further from different perspectives in Chapters 6 and 14. Nevertheless, nowadays, being an imitation infers that at best the item is inferior and at worst something setting out to deceive: a fake. Thesauri are full of pejorative synonyms for plastics such as 'ersatz', 'false', 'phony', 'pseudo' and 'unnatural' (Kipfer 2007, 630). In some parts of the world, two-faced people are called 'plastic' or even 'Tupperware'. For example, a Filipino migrant in Canada referred to her employer as 'Mrs Tupperware' behind her back because she was 'so *plastik*—she says I can have a writing desk, time off, overtime but none arrives' (McKay et al. 2015, 179). There are convincing arguments that can be made in favour of imitation, for example, if, as in the case of the combs, it results in the protection of endangered species and opens up perceived luxury goods to a wider public. However, when it comes to plastic flowers it is harder to find redeeming arguments. Perhaps it is for this reason that in spite of the long-lived and ongoing trade in plastic flowers, they are scarcely mentioned in books on plastics and are low down the pecking order even in accounts of the history and craft of artificial flowers. Kirsten Hardie addresses this neglect in this volume, with reference to Bourdieu's writing on taste, through an appraisal of how plastic flowers are valued by different people in different contexts, re-instating them as artefacts worthy of research and capable of meaning (Bourdieu 2010).

The imitative achievements of plastics have tended to disguise their innovative value. For example, celluloid film made available to the public by George Eastman, the inventor of the trade name Kodak, in 1889, transformed photography. It turned it from a gentleman's art requiring fragile and expensive paraphernalia into an egalitarian medium that gave rise to the family snap and provided a mass public with a means of recording memories (Fenichell 1996, 53–63). The addition of perforations along the sides of the film created cinematic film. It was for almost a century the vehicle that brought a wide range of entertainment within almost everyone's reach within the developed world. Plastics have also made music of our choice available at home, initially through shellac records, then through vinyl and more recently through polycarbonate CDs. Even now that digital recordings are taking the place of plastics, intriguingly, vinyl is making a comeback. Sales of vinyl reached a 25-year high in 2016 (Ellis-Petersen 2017).

Plastics increasingly became less valued for what they could look like and more valued for what they could do or make happen. An application that it is claimed influenced the direction of history was the result of the dielectric properties of polyethylene, a British discovery, which facilitated airborne radar. It is said to have given British pilots the critical edge during World War II and thus to have contributed to the war's outcome (Fenichell 1996, 202). Yarsley and Couzens, writing at that time, took this argument a step further. They claimed that plastics enabled countries that lacked such essential natural resources as coal and metal to make up for these deficiencies and that, thus, plastics were instrumental in modifying the balance of industrial power (Yarsley and Couzens 1941, 110–112). This is a subject explored in depth by Eli Rubin in respect of East Germany (2009). Certainly, plastics opened up possibilities on a global scale.

In this volume, textiles curator Tone Rasch provides such a case study of the introduction of viscose fibres to Norway's existing textile industry as a political act between the world wars. These fibres were extraordinarily alien to the largely agrarian character of Norway's economy on account both of their foreign origin and synthetic nature. She demonstrates how they were marketed differently to men and women. Women's magazines presented them as the height of fashion whereas newspapers, mainly read by men, stressed their potential value to the country's industrial strength. Certainly, as stated earlier, viscose fabrics had at this date a number of drawbacks. Thus, their successful introduction will have relied principally on the contribution they made to the Norwegian economy.

The part played by plastics during World War II raised expectations in terms of what plastics could deliver but it took about two decades before these expectations began to be fully realised. A contributing factor to their acceptance was the development of new kinds of plastics known as copolymers, formed by a reaction between two or more monomers, which increase the particular plastic's desirable properties. An example is styrene acrylonitrile (SAN). It is made up of approximately three-quarters styrene and one-quarter acrylonitrile, which is stronger, more rigid, and has better chemical resistance than styrene alone. It was used to manufacture the *EKCO Nova* range of stackable crockery designed by David Harman Powell, which won the Duke of Edinburgh Award for elegant design in 1968. Its reception provides the focus for Maria Georgaki's examination in this volume of the disparity between modernist design values and public proclivity.

The context of Georgaki's discussion is a collection formed with the purpose of educating the public about 'good design'. Although her focus is a group of plastics artefacts, representation of plastics within the collection is relatively sparse. Deborah Cane and Brenda Keneghan develop this theme through an exploration of the reception of plastics in museums and galleries, specifically the Victoria and Albert Museum and Birmingham Museums Trust. In 2007, ICOM defined a museum as an institution 'which acquires, conserves, researches, communicates and exhibits the tangible and intangible heritage of humanity and its environment...' (ICOM n.d.). Therefore, given the endemic infiltration of plastics in our heritage, I was surprised to be asked in 2008 in my role as curator of the Museum of Design in Plastics, to contribute to a series of workshops on 'Extreme Collecting', defined as 'the process of collecting that challenges the bounds of normally acceptable practice' (University College London and British Museum 2008; Lambert 2012). The fact of my invitation suggests that plastics are in some way seen as outside the traditional museum canon. This proposition is supported by Cane and Keneghan. They argue that plastics things have entered museums in spite, rather than because, of their material and explore how this may influence how they are valued by both museum professionals and the visiting public.

Cane and Keneghan highlight an artwork by the leading Benin artist, Romuald Hauzoumè. It represents a Benin street trader's cart with a wide range of merchandise, much of it made of plastics, which although still functioning has been discarded. His standpoint is, 'I send back to the West that which belongs to them, that is to say, the refuse of consumer society that invades us everyday' (Pigozzi n.d.). This artwork was purchased by the Art Fund for £82,000 (Art Fund 2013, 29). That sum, however, was not predicated on the value of the work's components but rather on the value attached to the work of a leading twentieth-century artist.

Nonetheless, there is a thriving collectors' market in vintage plastics, which is the subject explored by academic and avid plastics collector, Gerson Lessa. He focuses on products made of cast phenolic resins describing their visual and tactile qualities as similar to those of gems. He examines what it is that makes certain plastics artefacts rather than others desirable, concentrating on their materiality and processual qualities. He concludes that favourable perception grows in direct proportion to the mass of the object and the extent to which it is hand-finished. Such pieces can sell for figures of four and sometimes five digits (although not six) and have led to the creation of a counterfeit market known as 'fakelite'.

However, much more common than these treasured items are the single-use items, such as lighters, water bottles, coffee cups, food trays and infamous plastic bags, all of which are part of everyday life. The concept of

'Throwaway Living' appears to date back to an article, thus titled, in a 1955 issue of *Time* magazine. This article, subtitled 'Disposable items cut down household chores', claims that they save forty hours of housework and shows a couple and a child leaping gleefully as single-use products rain down on them. However, although some of the products may well have been made of plastics, it is paper and foil rather than plastics products that are mentioned in the text (Anonymous 1955, 43–44). Thus, plastics did not, on their own, invent the 'single-use' concept, but they have certainly fed what was probably in 1955 more of a fantasy than a reality and sixty-five years on are leading to ever more disastrous consequences.

As is widely documented, plastics debris in our oceans is now leading to horrific deaths for increasing numbers of wildlife, contamination of our food chain and modification of entire ecosystems with floating plastics transporting flora and fauna to places they would never reach under their own steam. However, it is not the fault of the plastics or the products that they end up in such inappropriate and damaging places but rather a result of our own careless actions. Each side of the exit from the local motorway services, close to where I live, is littered with the plastics packaging of consumed snacks thrown from car windows. Almost 90% of ocean plastics comes from ten rivers, two in Africa and eight in Asia, all from countries where public collection of waste is inadequate (Krauth et al. 2017). Recycled plastics have a monetary value and 87% of the recycled plastics collected by the EU and large quantities also from America and Japan have until recently been exported directly (or indirectly via Hong Kong) to China (Hofford 2017). However, since the start of 2018, China has banned the import of waste. It is to be hoped that this will lead to governments giving the development of a global strategy to manage plastics waste greater urgency.

Susan Mossman's chapter provides a detailed snapshot of the status quo and current deadlines that have been set in place to improve social responsibility towards plastics use and disposal at domestic, corporate and national levels. She posits that it may be the cheapness of plastics products that leads us to treat them with contempt. This proposition is supported by a conversation Freinkel had with a lighter enthusiast about throwaway Bic lighters: it is 'a pretty sad world when people would accept a plastic lighter as a thing they'd carry in their pocket … there's no esteem in it', which prompted her to reflect, 'If you can't reuse or repair an item, do you ever really own it? Do you ever develop the sense of pride and proprietorship that comes from maintaining an object in fine working order?' (Freinkel 2011, 138–139).

The sections in this volume are reciprocal. The chapters on *Plastics in Professional Plastics* provide different perspectives on related issues to the chapters on *Plastics in Societal Use*. In particular, Mossman's complements Bischof's chapter. Both clarify the importance of valuing plastics: one as a manufacturer and the other as a materials' scientist. From their different perspectives, they both stress the importance of using plastics responsibly and disposing of them responsibly. The most sustainable model for manufacture and consumption is one in which design, production, use and disposal result in a carbon neutral cradle-to-cradle existence, as envisaged by Braungart and McDonough (2002). Only if we succeed in making a circular plastics' economy a reality will this extraordinary, paradoxical, largely manmade family of materials live up to the expectations of their inventors and fully realise their innate value.

Notes

1. Cellulose nitrate and celluloid are the same material: the first being the English term and the latter the American.
2. A name used by Parkes, prior to Parkesine for his invention.
3. In some of these cases, plastics alone are used and in others, plastics provide vital insulation for metal parts.

Bibliography

American Crafts Council. 1968. *PLASTIC as Plastic*. New York: Museum of Contemporary Crafts.
Anonymous. 1943. Plastics: War Makes Gimcrack Industry into Sober Producer of Prime Materials. *Life*, May 3. https://books.google.co.uk/books?id=500EAA AAMBAJ&pg=PA65&lpg=PA65&dq=War+makes+Gimcrack+Industry+into+ sober+producer+of+prime+materials&source=bl&ots=akFXWIuIYo&sig=QkE lp2ExSMd-gh6fZUcBGtw0r0M&hl=en&sa=X&ved=0ahUKEwifwZG4k9XX AhXhBcAKHe6BAEsQ6AEIJjAA#v=onepage&q=War%20makes%20 Gimcrack%20Industry%20into%20sober%20producer%20of%20prime%20 materials&f=false. Accessed 23 January 2017.
———. 1955. Throwaway Living: Disposable Items Cut Down Household Chores. *Time*, August 1. https://books.google.it/books?id=xlYEAAAAMBAJ &printsec=frontcover#v=onepage&q&f=false. Accessed 20 January 2020.
Art Fund. A Year in Our Life...2012/13. 2013. https://www.artfund.org/ assets/news/2013/06/annual-report/art-fund-annual-review-2012-13.pdf. Accessed 18 May 2020.

Atkinson, Tracy. 1969. *A Plastic Presence*. Milwaukee: Milwaukee Art Center.

Barthes, Roland. 2000. Plastic. In *Mythologies*, trans. Annette Lavers, 97–99. London: Vintage. (Original French Edition 1957).

BBC News Channel. 2003. South Africa Bans Plastic Bags. May 9. http://news.bbc.co.uk/1/hi/world/africa/3013419.stm. Accessed 14 September 2017.

Beinecke Rare Book and Manuscript Library. n.d. Alexander Calder's Entry which Received First Prize among the 250 Entries. Yale University: Beinecke Digital Collections. https://brbl-dl.library.yale.edu/vufind/Record/3529872. Accessed 30 October 2017.

Bennett, Jane. 2010. *Vibrant Matter*. Durham, NC: Duke University Press.

Big Fat Bag. n.d. List by Country; Bag Charges, Taxes and Bans. http://www.bigfatbags.co.uk/bans-taxes-charges-plastic-bags/. Accessed 4 September 2017.

Bourdieu, Pierre. 2010. *Distinction: A Social Critique of the Judgement of Taste*. London and New York: Routledge Classics.

Braungart, Michael, and William McDonough. 2002. *Cradle to Cradle: Remaking the Way We Make Things*. New York: Farrar, Strauss and Giroux.

Cascini, G., and P. Rissone. 2004. Plastics Design: Integrating TRIZ Creativity and Semantic Knowledge Portals. *Journal of Engineering Design* 15 (4): 405–424.

Cole, Jack. 1941. Plastic Man. Police Comics, 1: August.

Conran, Sebastian. 2005. V=[D+Q+B]/P [Creating Value]. In *The Art of Plastics Design*, 2–4. Shrewsbury: Rapra Technology Ltd.

Disney, Walt. 1944. Donald Duck the Plastics Inventor. https://www.youtube.com/watch?v=ZnOcmrlFng0. Accessed 23 November 2017.

Dubois, J. Harry. 1972. *Plastics History*. Boston: Cahners Books.

Edwards, Chris, and Jonna Meyhoff Fry. 2011. *Life Cycle Assessment of Supermarket Carrier Bags: A Review of the Bags Available in 2006*. Environment Agency. https://www.gov.uk/government/uploads/system/uploads/attachment_data/file/291023/scho0711buan-e-e.pdf. Accessed 18 May 2020.

Ellis-Petersen, Hannah. 2017. Record Sales: Vinyl Hits 25-Year High. *The Guardian*, January 3. https://www.theguardian.com/music/2017/jan/03/record-sales-vinyl-hits-25-year-high-and-outstrips-streaming. Accessed 20 January 2020.

Espacenet Patent Search. n.d. Bag with Handle of Weldable Plastic Material. https://worldwide.espacenet.com/publicationDetails/originalDocument?CC=US&NR=3180557&KC=&FT=E. Accessed 27 August 2017.

Fallan, Kjetil, ed. 2019. *The Culture of Nature in the History of Design*. London: Routledge.

Fenichell, Stephen. 1996. *Plastic: The Making of a Synthetic Century*. New York: HarperCollins.

Freinkel, Susan. 2011. *Plastic: A Toxic Love Story*. Boston and New York: Houghton Mifflin Harcourt.

Friedel, Robert. 1984. The Plastics Man; Leo Baekeland's Conquest of an Unyielding, Brittle Resin Led to Nylons and Tupperware. *Science 84* 5: 49. Gale Academic OneFile.

———. 1987. The First Plastic. *American Heritage of Invention and Technology* 2 (2): 18.

Gabrys, Jennifer, Gay Hawkins, and Mike Michael, eds. 2013. *Accumulation: The Material Politics of Plastics.* London and New York: Routledge.

Google Patents. n.d. Patents: Improvements in the Manufacture of Combs from Celluloid. http://www.google.com/patents/US199909. Accessed 15 October 2017.

Gough, Marion. 1947. The Truth About Plastics. *House Beautiful,* October.

Hammer, Martin, and Christina Lodder. 2000. *Constructing Modernity: The Art & Career of Naum Gabo.* Newhaven and London: Yale University Press.

Hawkins, Gay. 2013. Made to Be Wasted: PET and the Topologies of Disposability. In *Accumulation: The Material Politics of Plastics,* ed. Jennifer Gabrys et al., 49–67. London and New York: Routledge.

Hodge, G. Stuart. 1968. *Made of Plastic.* Flint: Flint Institute of Art.

Hofford, Alex. 2017. China Bans Foreign Waste – But What Will Happen to the World's Recycling? *The Conversation,* October 20. http://theconversation.com/china-bans-foreign-waste-but-what-will-happen-to-the-worlds-recycling-85924. Accessed 20 January 2020.

ICOM. n.d. https://icom.museum/en/resources/standards-guidelines/museum-definition/. Accessed 18 May 2020.

INCPEN. 2008. Irish Carrier Bag Tax. http://www.incpen.org/displayarticle.asp?a=55&c=2. Accessed 4 September 2017.

Katz, Sylvia. 1978. *Plastics: Designs and Materials.* London: Studio Vista.

Keneghan, Brenda, and Louise Egan, eds. 2009. *Plastics: Looking at the Future and Learning from the Past.* London: Archetype Publications.

Kipfer, Barbara Ann, ed. 2007. *Roget's 21st Century Thesaurus.* Bantam Dell: New York.

Koestler, Robert J. 2017. *The Age of Plastic: Ingenuity and Responsibility: Proceedings of the 2012 MCI Symposium.* Washington, DC: Smithsonian Institution Scholarly Press.

Krauth, Tobias, Christian Schmidt, and Stephan Wagner. 2017. Export of Plastic Debris by Rivers into the Sea. *Environmental Science and Technology.* https://pubs.acs.org/doi/10.1021/acs.est.7b02368#. Accessed 18 May 2020.

Lambert, Susan. 2012. Plastics – Why Not? A Perspective from the Museum of Design. In *Extreme Collecting, Challenging Practices for 21st Century Museums,* ed. Graeme Were and J.C.H. King, 168–180. New York and Oxford: Bhergahn Books.

Latour, Bruno. 2004. *Politics of Nature: How to Bring The Sciences into Democracy.* Cambridge: Harvard University Press.

Mailer, Norman. 1988. *Conversations with Norman Mailer.* Ed. J. Michael Lennon. Jackson: University of Mississippi Press.

Manzini, Ezio. 1989. *The Material of Invention.* Cambridge, MA: The MIT Press.

———. 1992. Plastics and the Challenge of Quality. In *Change Design.* http://www.changedesign.org/Resources/Manzini/Manuscripts/Plastics%20and%20the%20Challenge%20of%20Quality.pdf. Accessed 1 September 2017.

McKay, Deirdre, et al. 2015. Subversive Plasticity: Materials Histories and Cultural Categories in the Philippines. In *The Social Life of Materials: Studies in Material and Society,* ed. Adam Drazin and Susanned Küchler, 175–192. London: Bloomsbury Academic.

Meikle, Jeffrey L. 1995. *American Plastic, a Cultural History.* New Brunswick and London: Rutgers University Press.

Meyer, Richard. 2013. *What Was Contemporary Art?* Cambridge, MA: The MIT Press.

Mieskowski, Katharine. 2007. Plastic Bags Are Killing Us. https://www.salon.com/2007/08/10/plastic_bags/. Accessed 4 September 2017.

Miller, Daniel. n.d. *The Uses of Value.* https://www.ucl.ac.uk/anthropology/people/academic-and-teaching-staff/daniel-miller/uses-value. Accessed 13 January 2018.

Milward, Bob. 2000. Theory of Value. In *Marxian Political Economy: Theory, History and Contemporary Relevance,* ed. Bob Milward, 28–37. London: Palgrave Macmillan.

Moore, Henry. 1934. Statement for Unit One. In *Unit One: The Modern Movement in English Architecture,* ed. Herbert Read, 29–30. https://www.tate.org.uk/art/research-publications/henry-moore/henry-moore-statement-for-unit-one-r1175898. Accessed 23 October 2017.

———. 1970. Some Notes of Space and Form in Sculpture. In *Henry Moore: Sculptural Process and Public Identity.* https://www.tate.org.uk/art/research-publications/henry-moore/henry-moore-some-notes-of-space-and-form-in-sculpture-r1145426. Accessed 23 October 2017.

Morris, William. 1892. The Influence of Building Materials on Architecture. In *Century Guild Hobby Horse.* https://www.marxists.org/archive/morris/works/1891/building.htm. Accessed 23 October 2017.

Mosley, Leonard. 1980. *Blood Relations: The Rise and Fall of the Du Ponts of Delaware.* New York: Athenaeum.

Museum of Design in Plastics. 2019. http://museumofdesigninplastics.blogspot.com/search?q=e-waste. Accessed 6 May 2020.

———. n.d. http://www.modip.ac.uk/artefact/phsl-x12-0. Accessed 13 January 2018.

Ndiaye, Pap A. 2007. *Nylon and Bombs: DuPont and the March of Modern America.* Baltimore, MD: John Hopkins University Press.

Onyanga-Omara, Jane. 2013. Plastic Bag Backlash Gains Momentum. *BBC News*, 14 September. http://www.bbc.co.uk/news/uk-24090603. Accessed 4 September 2017.

Painter, Paul C., and Michael M. Coleman. 2009. *Essentials of Polymer Science and Engineering*. Lancaster, PA: DEStech Publications, Inc.

Pigozzi, Jean. n.d. Contemporary African Art Collection: Romuald Hazoumè. http://caacart.com/pigozzi-artist.php?i=Hazoume-Romuald&bio=en&m=35. Accessed 15 January 2018.

Plastic Materials. 1933. *Nature* 131, 510. https://doi.org/10.1038/131510a0. https://www.nature.com/articles/131510a0#citeas. Accessed 30 November 2019.

Plastics Historical Society. n.d. Parkesine. http://plastiquarian.com/?page_id=14375. Accessed 23 October 2017.

Quye, Anita, and Colin Williamson, eds. 1999. *Plastics Collecting and Conserving*. Edinburgh: NMS Publishing Limited.

Rubin, Eli. 2009. *Synthetic Socialism: Plastics and the Dictatorship in the German Democratic Republic*. Chapel Hill: University of North Carolina Press.

Ruskin, John. 1849. *Seven Lamps of Architecture*. Boston: Dana Estes & Company. https://www.ajhw.co.uk/books/book435/book435. Accessed 23 October 2017.

Scottish Plastics and Rubber Association. 2013. Plastic Bag Tax Revisited. https://www.spra.org.uk/2013/09/14/plastic-bag-tax-revisited/. Accessed 20 January 2020.

Semuels, Alana. 2019. The World Has an E-Waste Probsslem. *Time*, May 23. https://time.com/5594380/world-electronic-waste-problem/. Accessed 18 May 2020.

Shashoua, Yvonne. 2008. *Conservation of Plastics, Materials Science, Degradation and Preservation*. Amsterdam: Elsevier.

Slosson, Edwin. 1919. *Creative Chemistry Descriptive of Recent Achievements in the Chemical Industries*. New York: The Century Co. http://www.ajhw.co.uk/books/book263/book263.html. Accessed 23 October 2017.

Smith, Adam. 2009. Updated 2019. An Inquiry into the Nature and Sources of the Wealth of Nations, 1: 1V. http://www.gutenberg.org/files/3300/3300-h/3300-h.htm. Accessed 20 January 2020.

Sparke, Penny, ed. 1990. *The Plastics Age: from Modernity to Post-Modernity*. London: Victoria and Albert Museum.

Thomas, Kevin. 2005. A Far-Out Night with Andy Warhol, The Los Angeles Times, May 1966. In *All Yesterdays' Parties, the Velvet Underground in Print 1966–1971*, ed. Clinton Heylin. Massachusetts: Da Capo Press. https://books.google.co.uk/books?id=MhpKDgAAQBAJ&pg=PT28&lpg=PT28&dq=Kevin+Thomas+a+far+out+night+with+andy+warhol&source=bl&ots=gmcRXmCzKT&sig=ACfU3

U3Tv2mWbQ5X6GCrwYocY99Xg3euog&hl=en&sa=X&ved=2ahUKEwix16
LGz7_pAhXDlFwKHa-2Bb4Q6AEwEnoECAsQAQ#v=onepage&q=Kevin%20
Thomas%20a%20far%20out%20night%20with%20andy%20warhol&f=false.
Accessed 19 May 2020.

Thompson, Don. 2008. The Rehabilitation of Eel O'Brian. http://nurgh.
blogspot.co.uk/2008/09/rehabilitation-of-eel-obrian-by-don.html. Accessed
20 January 2020.

University College London and British Museum. 2008. Extreme Collecting.
http://www.ucl.ac.uk/extreme-collecting/. Accessed 20 January 2020.

V&A. n.d. Search the Collections: Tulip. http://collections.vam.ac.uk/item/
O144560/tulip-armchair-saarinen-eero/. Accessed 13 November 2017.

World Health Organisation. n.d. Drinking Water. http://www.who.int/water_
sanitation_health/monitoring/water.pdf. Accessed 17 May 2020.

Yarsley, V.E., and E.G. Couzens. 1941. *Plastics*. London: Penguin.

PART I

Plastics in Professional Practice

CHAPTER 2

The Material Consciousness of Plastics

Russell Gagg

But the price to be paid for this success is that plastic … hardly exists as substance. Its reality is a negative one: neither hard nor deep, it must be content with a 'substantial' attribute which is neutral in spite of its utilitarian advantages … In the hierarchy of the major poetic substances, it figures as a disgraced material, lost between the effusiveness of rubber and the flat hardness of metal, it embodies none of the genuine produce of the mineral world…. (Barthes 2000, 98)

Written in 1957, the essay refers to Barthes observing those waiting in line to see an exhibition of plastics products, 'the stuff of alchemy' as a machine transmuted granular, raw material into everyday objects; the public bearing 'witness [to] the accomplishment of the magical operation par excellence'. Such an opening suggests a sense of the other-worldly, the wondrous, almost religious, the fascination of the populace with something they cannot hope to grasp. Barthes is suggesting that plastics, in his contemporary period already of the day-to-day, is a material and process of magic and myth and by being so has a meaning beyond the immediate and obvious. He even jokes that the names given to the variety of materials, 'Polystyrene, Polyvinyl, Polyethylene', are all those of ancient 'Greek

R. Gagg (✉)
Arts University Bournemouth, Poole, UK
e-mail: rgagg@aub.ac.uk

S. Lambert (ed.), *Provocative Plastics*,
https://doi.org/10.1007/978-3-030-55882-6_2

shepherds' (Barthes 2000, 97), a tangential connection with the classical world bestowing an enhanced provenance, commensurate with the finest art perhaps, on the spectacle that awaits those at the end of the queue. However in-thrall to this alchemy those he was observing might have been, Barthes was far from being in awe of what he saw. His initial exaltation of the miracle is short-lived and within a few lines the language changes from amazement to that of distaste and derision—for the material, its products and perhaps those that covet them.

> The fashion for plastic highlights an evolution in the myth of 'imitation' materials. It is well known that their use is historically bourgeois in origin (the first vestimentary postiches date back to the rise of capitalism). But until now imitation materials have always indicated pretension, they belonged to the world of appearances, not to that of actual use; they aimed at reproducing cheaply the rarest substances, diamonds, silk, feathers, furs, silver, all the luxurious brilliance of the world. Plastic has climbed down, it is a household material. It is the first magical material which consents to be prosaic. (Barthes 2000, 98)

What could be more dismissive of this magic than to consider it so commonplace and lacking in poetry?

Writing about the representation of plastics in the early twentieth century, Jeffrey Meikle also refers to this same, early, fascination with the material as 'the appeal of alchemical imagery: out of chemicals, powders, compounds, raw materials beyond the ken of ordinary folk' (Meikle 1992). Meikle refers to attempts made by the early pioneers of plastics to link the materials, their origins, processes and products, to those of more noble times and endeavours. Even the editors of *Fortune* magazine in 1940 'seemed uncertain as to how to present these new materials to their readership of business executives, whether to portray plastic as an extension of natural materials, a benign growth upon the body of nature, or as an intoxicating, even explosive disruption of the natural order and of human reason as historically understood' (Meikle 1992).

Fortune's efforts resulted in a series of commissions, explored by Meikle, that despite their best intentions, did little to resolve the confusion. The first being a map of 'an imaginary continent called Synthetica' which linked an image of exotic lands, reminiscent perhaps of South America or Africa, those of former mineral extraction and imperial wealth, with the materials of the new age. Countries such as Cellulose, Vinyl and

Urea contained within them geographical features: the Great Acetylene river, the Crystal Hills mountain range and the Acetic Acid Lake (Fig. 2.1).

> The map of Synthetica firmly rooted plastic in the extractive materials culture of the past. The conceit of a map implied an odd denial of organic chemistry's promise to free the human race from reliance on geographic accidents of scarcity and supply. The very outline of the continent suggested the tropics, civilization's colonial preserves exploited for ivory, rubber, gutta percha, copal, and other natural resins and substances. (Meikle 1992)

This attempt to connect plastics with a more valued provenance also took the form, in the same article, of '*An American Dream of Venus*', a pop-art photomontage with an acrylic, armless, female torso at its centre, most likely referencing her priceless ancestor in the Louvre. Abundantly accessorised with a wide selection of daily objects (including sunglasses, cameras and a pink toothbrush), the illustration 'implied radical divorce from the limitations of a drab pre-plastic existence … the limitless horizons, strange juxtapositions, endless products of this new world in process of becoming' (Meikle 1992).

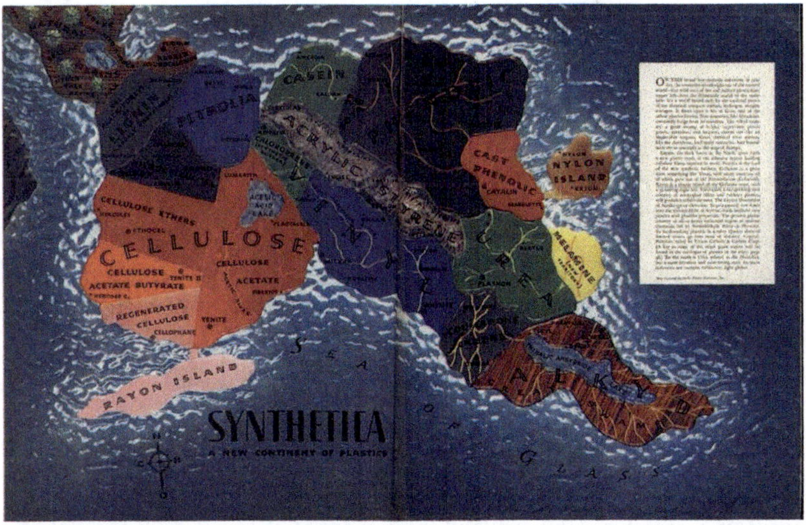

Fig. 2.1 Map of *Synthetica, Fortune,* October 1940

It is clear, and receives balanced commentary from Meikle, that in the early part of the twentieth century, plastics were seen as leading a new utopian vision for economic growth after the depression of the pre-war and austerity of the post-war years. I would suggest that what is also clear from Meikle's research is that throughout the 1920s to the 1950s (the period of his article), there were a number of critics, those today, perhaps, we would call the liberal elite, who echoed Barthes' view on the explosion of mass-produced articles of dubious taste: 'As early as 1936, the author of Fortune's first survey of plastic had recoiled from bright-coloured Catalin cast phenolic because 'nothing else can match it in potential gaudiness'. The reader was asked to 'shudder for the future because the stuff can be made in any colour that suits the designer' (Meikle 1992).

Norman Mailer, in his column 'The Big Bite' for *Esquire* magazine of May 1963, expressed his obvious distaste with the new materials: 'if everywhere we are assaulted by the faceless plastic surfaces of everything which has been built in America since the war, that new architecture of giant weeds and giant boxes, of children's colors [*sic*] on billboards and jagged electric signs. Like the metastases of cancer cells, the plastic shacks ... proliferate year by year until they are close to covering the highways of America with a new country which is laid over the old...' (Mailer 1963).

Thomas Hine, the American writer on history, culture and design, in *Populuxe*, his history of the era, observed that 'The decade from 1954 to 1964 was one of history's great shopping sprees, as many Americans went on a baroque bender... Products were available in a lurid rainbow of colours and a steadily changing array of styles. Commonplace objects took extraordinary form, and the novel and exotic quickly turned commonplace' (Hine 1986, 3).

What did these observers of twentieth-century culture have in common? They were writing in a period of transition, a time when, either through economic necessity or technological development, or both, notions of materials and making, of craft, were changing. Materials that had been tangible, elemental, real, understood, were crafted into objects that themselves were understood and retained the inherent qualities and character of the materials from which they had been made. What was made was informed by and intrinsically derived from the nature of the material used. We know now that plastics changed all this irrevocably. Leaving aside a certain moral high ground on which these authors seem to position themselves (the availability of more to the masses disturbs them as much as the objects do) perhaps they could foresee what was coming?

Were they also un-nerved by magical processes that they did not quite understand? Was their unfavourable comparison of plastics with a more worthy classical tradition an attempt at rationalising their distrust?

Their writing contextualises plastics and its products both historically and today. I would suggest that the ubiquity of this material has not diminished its mystery: 'ordinary folk' having, probably, little more knowledge now of where it comes from and how it is made than those waiting in line in 1957. If anything, the knowledge that we do have focuses not on the material itself but its impact. Press coverage of ecological and biological damage is comprehensive and increasing. Scientific analysis of the unseen dangers posed by plastics is presented across mainstream media on almost a daily basis. Our governments respond to our concerns with fine words, strategies and policies that promise action—if only we could decide what to do.

However, the purpose of this chapter is not to pretend that these debates concerning consumption and contamination are not real and pressing in our contemporary culture. Instead, by raising such questions through looking at remarkable design, I aim to offer an alternative argument for a new appreciation of a much-maligned material.

Central to this alternative view is the consideration of plastics as contemporary craft, surely anathema to Barthes, Hine, Mailer and, most recently, Meikle, but a view I hope to justify through recourse to the work of an equally prestigious writer, that of Richard Sennett.

In the summer of 2015, we were lucky enough to spend a few days helping friends build a pizza oven in their back garden. First, we had to 'puddle' the clay: mixing the wet clay with sand in order to make it more workable and structurally sound. This was a group activity with each person using their bodyweight to mix the materials together. As we rotated, and then as the clay was turned, all the constituents were gradually combined.

As well as being enjoyable, the spectacle of two families on the pavement, treading clay, caused much comment from passers-by. What was interesting was that all the comments were positive and good-humoured; it seemed that anyone who watched us (including the mounted police who happened to ride past) found some basic connection with such a fundamental process. Most touching of all, on our last round of mixing, a very elderly woman stood next to us, watched, and started to cry gently. When we stopped, she smiled and commented that seeing us had brought back memories of her making bricks as a young girl, with her sisters, parents and grandparents, all of whom were now dead.

From this mound of mixed clay, smaller handfuls were taken and 'wedged' or kneaded in order to form bricks; the kneading was essential to make sure that no air pockets were present in the clay (which would cause the brick to explode). This is exactly the same technique that any potter uses. The bricks were then stacked and pressed against the dome of clay (which had been formed around a sand mound) again making sure that there were no gaps in which air would be trapped. Finally a slurry, just clay and water, was applied to the finished dome to give a smooth covering and to fill any remaining gaps (Fig. 2.2).

From this point, over several weeks, progressively larger and hotter fires would need to be lit inside the oven in order to dry the clay walls which, when finished, were some 20 cm thick.

It was only when I looked back on the three days that I considered what had actually taken place and how alien this was to our everyday life, how important it had been and how closely it connected with Sennett's central theme and the theme central to this chapter: material consciousness.

Fundamental materials had been taken (earth and water), combined, worked and changed, *metamorphosised*, to create a new form; those forms had been added to others to create something new. It was the actions of those building and their understanding of the material that enabled this

Fig. 2.2 Applying the slurry to the dome of the pizza oven. Photo: Russell Gagg

creation to take place; the mark of our actions, *presence*, had become clear. The clay, the water, the puddling, the bricks, the construction, the finishing, the laughter and tears of those taking part and those watching all aroused human feeling, memory and emotion—all imprinted on a lump of inanimate earth: anthropomorphism.

Metamorphosis, presence and anthropomorphism are the basis for Richard Sennett's concept of material consciousness as written in his book, *The Craftsman*. He argues that we 'become particularly interested in the things we can change' and that material consciousness will only occur when a material is capable of change (metamorphosis), can record the mark of its maker (presence) and can be invested with human qualities (anthropomorphism) (Sennett 2009, 120).

In common with those writers cited at the start, Sennett also referred back to ancient times. Referring to Ovid's *The Metamorphoses*, 'My purpose is to tell of bodies which have been transformed into shapes of a different kind' (Sennett 2009, 123). Recounting the chronological creation of the world, *The Metamorphoses* tells of change, of transformations decreed by the gods that manifested their power, through nature into art. Physical reality was a never-ending recombination of the four elements: air, fire, earth and water; a shifting of shape and form seen to be wondrous and fearful. Working against what were seen to be the logical assumptions of the universe, such transformations were irrational: being of the emotions and the senses, the unexplainable, magical. This chapter argues that, contrary to popular perceptions, plastics engage with material consciousness as much as any of our ancient elements, that plastics are our contemporary magic. 'The real skill of the practitioner lies not in skilled concealment but in the skilled revelation of skilled concealment. Magic is efficacious not despite the trick but on account' of its exposure' (Taussig 1998, 222).

Working from my position that the design and use of plastics is worthy of craft, I will examine each of Sennett's criteria to show how the contemporary craftsman engages with the materiality and consciousness of plastics.

METAMORPHOSIS

It seems sensible to start by looking at the potter. Not only is *The Potters Tale* the opening section to Sennett's work but the act of treading and forming clay bricks was my inspiration for looking at this story. Sennett also uses the example of the potter to illustrate how the most fundamental of elements have been processed, progressively throughout history, to produce vessels which are decorated through the application of surface finishes and designs. The introduction of fire allowed the pot to be made hard and watertight and therefore useable; fire also allowed for the control of the surface finish. Changes and refinements in these processes over time have led to greater expressive possibilities and have increased the objects' economic value. The story of pottery has changed little over thousands of years, yet the robust nature of traditional forms and conventional techniques does still result in progressive work being produced by artists today. Such an artist is Grayson Perry whose work is of particular resonance for this article.

We have already seen how writers such as Barthes and publications like *Fortune* magazine found the need to reference plastics against an established tradition in order to make them understandable and acceptable (acceptable at least in the case of *Fortune*). Clay possesses a plasticity that enables the skilled potter to mould and decorate a seemingly endless variety of forms which, when fired, become fixed and intransmutable. Perry also follows this established tradition in the forms that he sculpts, such as Grecian-like urns and amphora. These familiar shapes initially give us a security of understanding: we think we know what we are looking at. However, Perry then presents finished works that challenge our expected perceptions by decorating them with a diversity of social, literary and autobiographical scenes of often disturbing and harrowing subject matter. In an interview with the Saatchi Gallery, Perry commented that 'People say, "why do you need to put sex, violence or politics or some kind of social commentary into my work?" Without it, it would be pottery. I think that crude melding of those two parts is what makes my work' (Perry n.d.).

What is interesting here in the context of my argument is that Perry draws a distinction between his work, which the interviewer suggests should be seen as fine art, and pottery which, presumably, is not. The debate around that distinction is not for this chapter; however, the taking of a base material and through the processes of change, in this case the decoration of surfaces which express the feelings, tensions and life of the artist, elevate the object to one of enhanced cultural, economic and social value.

This first aspect of metamorphosis allows us to parallel plastics where we have the fusion of earth's elements to create a new and unique compound which, through either further chemical processes or application of surface finishes, becomes decorated. Such decoration or refinements of form have often increased, or have perceived to have increased, the economic value of the object.

The artist Matthew Darbyshire also asks us to question our perceptions of craftsmanship, material and value. Again, in common with our writers, the subject of his work takes the classical form, a priceless artefact, such as the Roman marble statue of Doryphoros from Pompeii. He, however, renders it out of a relatively valueless material such as multiwall polycarbonate sheeting. What at first glance might appear to be a simplistic copy of an understood work of art is in reality a carefully crafted sculpture drawn and cut by the artist's own hand. What suggests standardisation, surely a defining characteristic of all things plastic, is anything but. 'The works are at once depersonalised but at the same time haunted by their maker. They are haunted through the imperfections and errors involved in their hand-made production, in the crafted rendering of the drawing and the rough edges of the fabrication. These imperfections question and deconstruct the fixed norms of mass production/diffusion of objects and images' (Honoré 2014). Perhaps as justification for the artist's approach, his work sells for a thousand times the cost of the raw plastics material.

Sennett notes that the ancient understanding of metamorphosis ultimately led to decay and he refers to Plato who proposed that although the physical decays, the idea of the thing endures. This has persisted in Western thought, argued Sennett, in the supposed superiority of the head (the philosopher) over the hand (the craftsman).

This is at odds with our general perception of plastics as being immutable and ever-lasting (which is central to an arguably more pressing ecological debate) but we could see this quality of many plastics as being 'magical'. So why do we see the material and its forms as being so disposable (after all, the raw hydrocarbons from which they are made are far from valueless and indeed dictate the global economy). Surely, one of the reasons we prize the sculptures that have been passed down to us from antiquity is precisely because they have lasted the tests of time, their very timelessness being intrinsic to our culture.

Sennett argues that 'the craftsman engage[s] in a continual dialogue with materials...' (Sennett 2009, 125) and also quotes the engineer and historian of technology, Henry Petroski, on the importance of 'salutary failure' as being crucial to the second aspect of metamorphosis which is where 'two or more unlike elements are joined'. The failure of a form inevitably leads to the analysis of details and small parts which then promote change and evolution of that form, 'this micro-address seems the sensible way to deal with failure or trial and error, and to Petroski the address bespeaks a healthy consciousness' (Sennett 2009, 126). The craftsman has to decide whether the new combination of elements will work best as a compound in which the whole becomes different from the original parts or as a mixture where the parts can be separated back to the original constituents. The parallel here is clear on the many, many refinements that have been made to the chemical constituents and manufacturing processes of plastics that have allowed us to produce almost limitless forms and finishes as well as specific materials for specific functions and specific conditions.

The final aspect of metamorphosis referred to is that of 'domain shift' (a term coined by Sennett). This is where the tool used for a process can be applied to another task; equally a guiding principle informing one practice can be applied to another activity.

To illustrate this domain shift, Sennett draws evolutionary connections between man-made processes that have helped underpin our civilisation. The orthogonal, interlocking threads of weaving, the weft and warp, ensured that the cloth produced was flexible, stable and strong. He argues that the same principle of interlocking elements was transferred to Greek ships in the first century BCE where mortise and tenon jointing of the timbers created a watertight, flexible and immensely strong hull enabling the Greeks to explore and colonise distant settlements. Finally, he remarks on the city plan, citing Selinus, Sicily, of the fourth century BCE, with its main thoroughfares arranged in a defined, right-angled, grid where corners were seen as opportunities for strength and intentional design and the city was compact, strong and tight as a result. Here, and on through history, 'urban fabric' became more than a casual metaphor.

Sennett's final comment on domain shift is arguably the most important, that of time: 'Like potting, these permutations in weaving occurred slowly, distilled by practice rather than dictated by theory. What endures, what does not decay, is the technique of focusing on the right angle. Domain shifts, when stated baldly, seem counterintuitive: at first glance it makes no sense to liken a ship to a cloth. But the craftsman's slow working through forges the logic and maintains the form. Many propositions that seem counterintuitive are not so; we just don't know their conditions yet. Plodding craft labour is a means to discover it' (Sennett 2009, 128). In the contemporary manufacture of plastics and plastics products, time and the plodding of craft may be measured in a different order. However, we see ancient processes such as moulding, casting and weaving as still fundamental to production; we see more recent technological achievements such as printing being appropriated—a domain shift—and leading to innovation.

As an educator, the use of 3D-printing technology enables our students to realise prototypes of their work quickly. The usual scenario is that the student thinks that the 3D-printer will provide, by the simple press of a button, a perfect, final representation of their idea. The reality is one of 'salutary failure' where many hours spent in consultation with the tutor will result in the generation of prototypes and components that, through the gradual refinement of different iterations of the computer model, eventually come together in a finished model.

For the professional designer, the use of cutting-edge technologies does not diminish an understanding of the passage of time and a continued dependency on traditional techniques. Demonstrating domain shift through one of our most basic anthropometric structures, Richard Liddle works with the cooled extrusion of reused HDPE (high-density polyethylene) to weave, by hand, his *Roughly Drawn Chair* of 2009 (Fig. 2.3).

The finished object is immediately recognisable as a chair but challenges our accepted understanding of what a chair should look like, being almost equally solid and void at the same time: 'Genuinely new design that starts with a blank piece of paper, a cup of coffee and a lot of headaches, well this takes time. It certainly isn't the easy path, but if at the end of the journey the new product has a market, it works simply, seamlessly and intuitively then the designer has done his job' (Liddle 2014).

Just as the early plastics were presented, erroneously or otherwise, as being drawn from the same heritage as our traditional materials, I would suggest that designers such as Richard Liddle and Francis Bitonti have felt the need to connect their highly innovative use of plastics materials and processes with the most everyday of objects in order to make their ideas accessible and acceptable. Bitonti's *Bristle Chair*, for example, combines seemingly individual branches of printed plastics (Fused Deposition Modeling 3D-printed filaments of ABS plastic to be precise) together to form a rigid, structural, mass; those that form the seat each ending in a supportive sphere. 'We are embracing a means of production that allows us to produce highly complex, customizable products. We are introducing formal complexity to mass produced objects, this have [sic] never been a possibility before. We are democratizing ornament' (Bitonti n.d.).

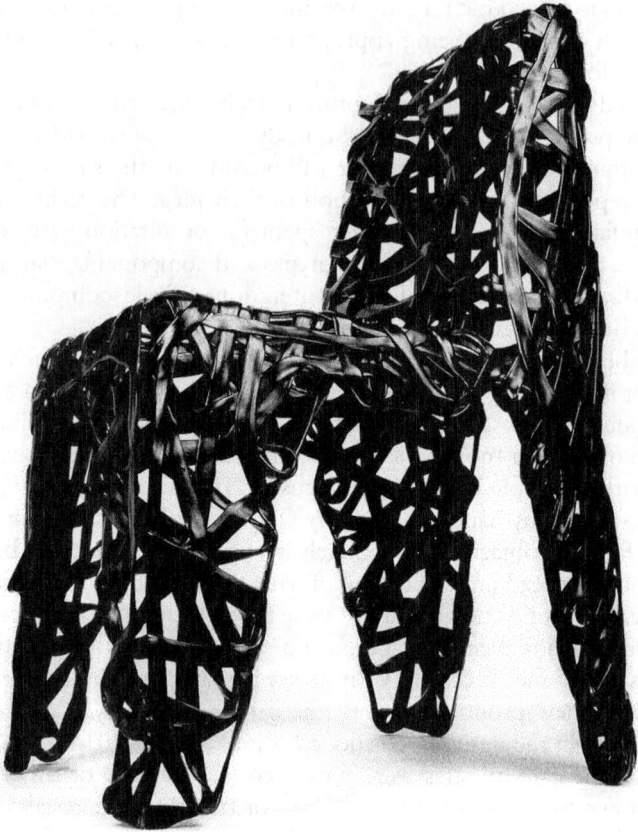

Fig. 2.3 *Roughly Drawn Chair* designed by Richard Liddle, 2009. Photo: Cohda Design

PRESENCE

Sennett draws an interesting distinction from ancient times between the mark of the Roman brick maker, typically a slave, who would leave an impression in the clay (usually a symbol denoting where the bricks were made) to record his presence. The mark made confirming that 'I exist', perhaps, in Sennett's view, the most important mark a slave could leave. The Greek potters, by contrast, began to sign the scenes painted on their pots; the signature could add economic value to the item.

The question arises of whether the maker/designer leaves their mark on the plastics artefact or do the necessities of manufacture preclude this? If we perceive plastics as being valueless, or at best disposable, then does the mark of the maker/designer matter in the context of the object that is produced? If the commercial manufacture of plastics products has to strike a balance between function and cost and the resulting objects are produced in such high numbers to make them economically viable then does it matter whether a name is known? Is the contemporary product/industrial designer any different from the Roman slave-brickmaker: both have produced many thousands of identical items for everyday use but at least the slave was able to signify 'I exist'.

A few designers have been able to leave their mark; they have taken full advantage of mass production methods and the ubiquity of plastics in the global market to produce the unique that is available to all. Precisely because of the global reach of plastics, their very 'disposability' (and yet, at the same time, indispensability), has allowed a few to achieve global recognition. Philippe Starck is certainly one such designer. His work, whilst not limited exclusively to the use of plastics, has arguably elevated the material to a status only previously enjoyed by those crafting the traditional. His long-term partnership with Italian designers and manufacturer Alessi has resulted in mass household objects that became the must-have products of a young, design-conscious generation. In Starck's own resumé, he comments on the use of plastics as an approach attempting to 'reconcile two opposing worlds: that of industrial high technology, that of mass production with that of the craftsman, reflecting original, unique and human excellence' (Starck cited in Wingfield, n.d.).

Starck clearly considers that his work should be valued as highly as that of the traditional craftsman. Certainly, one could argue that if a characteristic of what we perceive as being of value is something that lasts, then the immutability of plastics, that the *Dr Skud* Fly Swatter by Starck for Alessi will probably last some 400 years longer than his buildings (according to the US National Park Service Mote Marine Lab the average decomposition time for a plastics beverage bottle is 450 years) these are objects that, perhaps, we should treasure (Fig. 2.4).

Such 'brand extension', a term used by Sudjic in his book *The Language of Things* has allowed the designer to seek contemporary immortality not only through the ubiquity of their objects but also by their seeming indestructibility; the irony being that we no longer have to visit far-flung galleries to access the uniqueness of their craft, we merely click and collect (Sudjic 2009, 108).

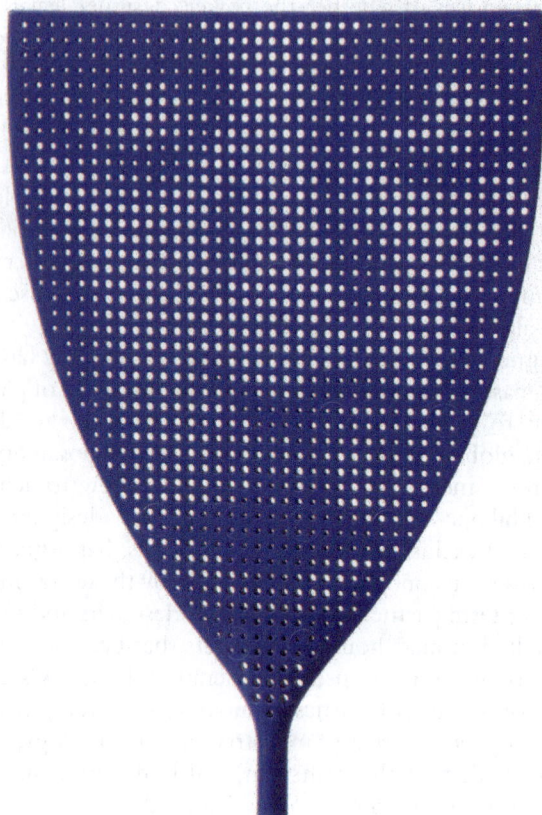

Fig. 2.4 *Mr Skud* fly swat (detail) design by Philippe Starck for Alessi, 1990s. Photo: MoDiP

ANTHROPOMORPHISM

Finally, I would like to consider Sennett's remaining argument for material consciousness that of anthropomorphism, the investing of human qualities in the non-human. Sennett suggests that 'The attribution of ethical human qualities—honesty, modesty, virtue—into materials does not aim at explanation; its purpose is to heighten our consciousness of the materials themselves and in this way to think about their value' (Sennett 2009, 137).

The clay, sand and water that we spent hours mixing with our feet, then forming into bricks with our hands, had no intrinsic, monetary value, but the engagement with the material and the processes that took place caused a depth of understanding and meaning to be invested in these lumps of inert mud that went way beyond their physical nature.

Plastics are probably one of the most highly engineered materials we have, the product of a short, but intense, period of chemical and manufacturing innovation that started at the turn of the nineteenth century but only really became of significance during the Depression years of the 1920s and 1930s.

When one picks up the clay pot or passes the hand against the marble statue, we feel the temperature of the surface, the marks of the tools used; we sense and probably have some understanding of the processes involved (the craft) in turning the raw materials into something of greater meaning yet despite the years of scientific experimentation and industrial development undertaken, it is questionable whether we have any sense or understanding of these achievements, or any concept of deeper meaning, as we load our plastic bowl into the dishwasher. Perhaps this is why companies such as Alessi feel the need to anthropomorphise our everyday products with cute forms and comical expressions.

Is it our metaphorical distance from these processes and our general inability to understand their foundation, beyond our ken as ordinary folk, or is it the reductivist approach to truly complex materials by companies such as Alessi (with their *Mr Suicide* bathplug, *Otto* dental floss dispenser and *Cico* egg cup) that inevitably leads us to consider plastics as fundamentally essential but ultimately disposable? Whilst this literal interpretation of anthropomorphism might seem to contradict my argument, I would suggest that the work of those that follow supports a deeper sense of attribution than is achieved by the quirkiness of Alessi.

LOOKING FORWARD

Their typical appearance was of seamless unitary creation. If they aped earlier forms or modes of construction, they did so only to enable users to integrate them into the known world. Plastic contributed to a 'black box' syndrome of ignorance about technological processes by enclosing them within smooth irreducible moulded forms of deceptive simplicity. A flawless surface, whether geometrically precise or sensuously flowing, implied an interior state of ideal perfection. Artefacts shaped by plastic during a self-conscious 'machine age' ironically indicated a willingness to ignore mechanical complexities, to abdicate responsibility for understanding or directing them, to assume that beneath these pristine surfaces everything was well under control. (Meikle 1992)

One of the most delightful arguments against this point is the work of Gangjian Cui who in 2015 exhibited at the Victoria and Albert Museum in *What Is Luxury?* Simply by inclusion in an exhibition of this title, his hand-made furniture challenged accepted perceptions but his premise goes further. Appropriating the language of the craftsman *Rise of the Plasticsmith* imagines the world of 2052 in which, due to the depletion of global petroleum resources, plastics have become an increasingly limited material. Prompted by his family history, he looks to his hometown of Daqing in Northern China, currently a large centre of plastics production, and its post-industrial future. He suggests how new skills will have to be developed for working with this now rare material.

Using a self-built, hand-operated, extrusion machine, Cui presses together the still malleable rods of acrylic to form joints to produce pieces of furniture. Arguably demonstrating the evidence and skills of craft, Cui's objects raise questions about value and our criteria for luxury materials and products. In his imagined world of scarcity (now not so difficult to imagine), materials and products on which we have become so dependent supplant those formerly considered to be precious. Even in this potential future, the position of plastics is still located in reference, and in deference, to perceptions of value and craft that have existed for millennia.

The paradox of Cui's scenario is that the role of plastics in our contemporary society has been one of economic revitalisation (especially during the inter-war years) and democratisation through enabling designed consumer goods to be commercially viable, socially desirable and globally abundant. 'Inexpensive plastic provided by organic chemists from out of common elements would foster true democratization of society by ending the strife generated by scarcity of raw materials and bring about a universal material abundance' (Meikle 1992).

In commenting on the rise of the Habitat retail brand, visualised and recognised by many through the marketing of plastics and their desirable forms, Deyan Sudjic writes, 'What really makes Conran stand out is the way he has been able to create a way of life that other people want to live. It is the world according to Conran, an attempt to make ordinary, banal places a little out of the ordinary' (Sudjic 2009, 108–109).

Perhaps our obsession with plastics comes from our need to make the banal a little less ordinary?

> The machine works away diligently and fills our bookcases with ill-printed volumes, its criterion is cheapness. Yet every cultured individual should feel ashamed of such material abundance. For on the one hand, ease of production leads to a diminished sense of responsibility, while on the other abundance leads to perfunctoriness. How many books do we genuinely make our own? And should one not possess these in the best paper, bound in splendid leather? Have we perhaps forgotten that the love with which a book has been printed, decorated and bound creates a completely different relationship between it and us, and that intercourse with beautiful things makes us beautiful? (Hoffman cited in Sudjic 2009, 106)

Originally written by Josef Hoffmann of the Viennese (Wiener) Werkstatte (1903–1932) in the early part of the twentieth century about the evils of industrial production and seen as a justification for the luxury of craftsmanship, the point is just as relevant today when faced with the strategy of abundant supply and easy disposal so readily marketed by contemporary global retailers. But not all those working with the ease and supply of plastics have adopted the same strategy. A few, perhaps too few, understand the potential for craftsmanship in this wonder material and have achieved a sense of luxury.

The latter part of the twentieth and early twenty-first centuries have seen advances in computer modelling and manufacturing techniques combine with greater sophistication in plastics and composite materials enabling designers to create meaningful and expressive objects and interiors that go beyond the cutesy kitsch of Alessi and Starck.

Ron Arad, Zaha Hadid and Karim Rashid stand as fine examples of the few designers working today who have realised the potential of plastics for what they are and expressed the fundamentals of the material: plasticity. Arad's chairs, such as the *Voido* rocking chair, designed in 2006 and manufactured from rotational-moulded polyethylene by Magis, sought, as other artists included here do, to redefine the idea of the chair. In capitalising on the qualities of plastics, Arad takes this much further in sculpting the object around the form of the sitter. The result is one of anthropomorphic sensuality and beauty that could only be achieved by an understanding of, and sensitivity to, the material being used. The architecture and interiors of the late Zaha Hadid, such as the ROCA London Gallery (2009–2011), are also characterised by a sensuous playfulness with composites. Plastic forms combine mathematical efficiency with seemingly impossible structural extremes; a counterintuitive experience of the ultimate in organic spaces created with the ultimate in inorganic materials. Playfulness shows off in the extreme kitsch of Rashid's interiors where bold colours and equally bold forms transform restaurants and retail into spaces rejecting pretension and extolling pure fun … just the sort of places that Barthes was dreading in 1957.

So, to take issue with Barthes, the success of plastics is that it very much exists as a substance; it exhibits form of the most poetic nature, is a material of grace and elegance and is, ultimately, genuine. Plastics, their inherent and defining plasticity, have enabled designers and artists such as Arad, Hadid and Rashid and all those written about here, the opportunity to create forms and spaces of such sensuality; artificial certainly but expressing an intimate understanding of the needs of the human psyche. This expression, by virtue of contemporary modelling, visualisation and manufacturing techniques has allowed the designer to express individuality and identity, to demonstrate contemporary metamorphosis, presence and anthropomorphism and exemplify craftsmanship. I would therefore argue that the material consciousness of plastics is as justified as that of 'the produce of the mineral world' and that plastics are the new magic.

BIBLIOGRAPHY

Barthes, Roland. 2000. Plastic. In *Mythologies*, Trans. Annette Lavers, 98. London: Vintage. (Original French Edition 1957).

Bitonti, Francis. n.d. Interview with Installation. Accessed February 2018. http://installationmag.com/francis-bitonti-the-algorithm-of-contemporary-design/.

Hine, Thomas. 1986. *Populuxe*. New York: Alfred A.Knopf.

Honoré, Vincent. 2014. Galerie Phillippe Jousse, May. Accessed January 2018. http://www.jousse-entreprise.com/en/contemporary-art/exhibitions/matthew-darbyshire-captcha.

Liddle, Richard. 2014. Interview with *Illumni*, February. Accessed January 2018. http://www.illumni.co/illumni-catch-richard-liddle-founder-cohda.

Mailer, Norman. 1963. The Big Bite. *Esquire Magazine*, May, 37. New York: Hearst Communications, Inc.

Meikle, Jeffrey. 1992. Into the Fourth Kingdom: Representations of Plastic Materials, 1920–1950. *Journal of Design History* 5 (3): 173–182. Accessed August 25, 2015. www.jstor.org/stable/1315836.

Perry, Grayson. n.d. Top of the Pots. Interview with the Saatchi Gallery. Accessed January 2018. http://www.saatchigallery.com/artists/grayson_perry_articles.htm.

Sennett, Richard. 2009. *The Craftsman*. London: Penguin.

Starck, Philippe. n.d. Philippe Starck Short Biography, Wingfield, Jonathan. Accessed February 2018. http://www.starck.com/en/about.

Sudjic, Deyan. 2009. *The Language of Things*. London: Penguin.

Taussig, Michael. 1998. Viscerality, Faith, and Skepticism: Another Theory of Magic. In *In Near Ruins: Cultural Theory at the End of the Century*, ed. Nicholas B. Dirks, 222. Minneapolis: University of Minnesota Press.

CHAPTER 3

Plastics' Value as a Sculptural Medium

Richard Hooper

Sculpture has traditionally been made of stone, wood and metal. This chapter explores instead the potential of plastics as a sculptural medium.

While the extraction, processing, distribution, use and disposal of materials have often been the subject of socio-economic and cultural debates, perhaps plastics, being mainly a synthetic industrial product, have elicited stronger voices than many. A chapter such as this, concerned with value in relation to materials, inevitably conjures up a variety of discourses that examine the extent to which sculptors are free actors in their field; free to choose the material for its sculptural value and/or free to comment on the positive or negative implications of the material. Some discourses may advance positive utopian technocratic visions in respect to the freedom of the sculptor. Others may offer more dystopian readings whereby industry in general, or sculptors themselves, are implicated in environmentally questionable and/or economically exploitative Marxist or Eisenhowerian Military-Industrial Complexes (Lillian Goldman Law Library 2008) and used wittingly (or unwittingly) in their associated marketing and public relations campaigns.

Notwithstanding such social constructivist approaches, this chapter explores instead the less charted territory of plastics' capacity to extended

R. Hooper (✉)
Liverpool Hope University, Liverpool, UK
e-mail: hooperr@hope.ac.uk

© The Author(s), under exclusive license to Springer Nature
Switzerland AG 2020
S. Lambert (ed.), *Provocative Plastics*,
https://doi.org/10.1007/978-3-030-55882-6_3

sculptural vocabulary, especially their potential for translucence and the impact on sculptural form of the relatively recently developed processes of computer aided design and manufacture (CAD/CAM), for which plastics are frequently the material of choice. This provides a frame of reference for an analysis of the author's investigation of the physical properties and semantic connotations of transparent acrylic in association with CAM/CAD processes in conceptualising in physical form the Christian notion of a tripartite deity. The influential twentieth-century British art historian, Herbert Read, described the pioneering translucent plastics sculpture of the Russian constructivist, Naum Gabo as 'hovering … between … the material and the immaterial [whereby their] axial system crystallise energy itself' (Read 1964, 112). He hailed one sculpture in particular as achieving 'the highest point ever reached by the aesthetic intuition of man' (Tate n.d.-a). The analysis of the author's work also proposes a rationale for such a hyperbolic reception of transparent plastics sculpture.

PLASTICS' CONTRIBUTION TO SCULPTURAL VOCABULARY

Plastics differ from other sculptural media in that they have no inherent characteristics: they can take any form, texture and colour. Thus they can be flexible or rigid, soft or hard, solid or ephemeral, colourful or colourless, translucent or opaque, and easy to work or intractable. These paradoxical characteristics provide an unusually wide range of possible formal outcomes. Some sculptors have even based the concept of their work around the versatility of the medium. This is the case with William Tucker's acrylic *Anabasis 1* of 1964, which consists of three interlocking sections. Each section is made up of a simple related cruciform shape; however, the lower one is angular with defined edges, the middle one is rounded as if inflated and the top one cut from a sheet to the plan of the others. The lower sections are opaque and white and the upper one transparent and yellow (Tate n.d.-b).

A consequence of plastics' versatility is their ability to imitate other materials which, as discussed from different perspectives in Chaps. 1, 6 and 9, has led to their denigration. This ability has, however, been turned to advantage by sculptors. The lifelikeness of the women, somewhere between real people and mannequins, in Allen Jones' 1969 painted fibreglass furniture pieces *Chair*, *Table* and *Hat Stand*, in which the figures contribute to the functions of the titles of the pieces, was central to their provocative intention (Wroe 2014). Works by super realists Duane Hanson and John De Andrea provide other examples of cast plastics, in this case polyester resin (sometimes reinforced with fibreglass) using silicone rubber moulds, to create life-size figures. These figurative sculptures represent real people, an example being Hanson's *Woman with a Purse* (1974) which for Lucie-Smith 'render[s] contemporary individuals in society in all their unattractiveness—but not without a certain sympathy for their fate in being what they are' (Lucie-Smith 1987, 68). Ron Mueck is also an artist who has used plastics for hyper-real effect. He typically alters perception of his figures by increasing their size, something the comparative cheapness of plastics as compared with many traditional sculptural materials, facilitates. For example, *A Girl* of 2006 created from fibreglass, silicon, polyurethane foam, and acrylic fibre with oil paint portrays a newborn baby, complete with umbilical cord, on a gigantic scale: just over five metres in length (National Galleries of Scotland n.d.). In each of these works, the frequently despised imitative capability of plastics is key to the realisation of the artist's vision.

Artists have also used plastics to imitate traditional media. Even Henry Moore became reconciled to the utility of plastics creating for example his bronze and travertine *Arch* sculptures (1963–1980) also in fibreglass versions resembling both materials (Serpentine Gallery 1978). However, some artists, such as Robert Morris, have used this imitative capacity as a means of reference that creates meaning. His 2015 installation *MOLTINGSEXOSKELETONSSHROUDS* was made by draping resin-soaked linen over mannequins, which were subsequently removed, leaving rigid, self-supporting shroud-like forms that movingly recall stone drapery. The installations in his 2017 exhibition *Boustrophedons* used the same methodology but with carbon fibre soaked in epoxy resin, an immensely strong yet lightweight material more commonly used in the manufacture of racing cars and aircraft. The result has been described as 'glossy and menacing, the fabric of choice for superheroes and villains in contemporary movies' (Karmel 2017, 4). But it has additional connotations. The material's similarity to bronze, although in no way reproducing it, will not have been accidental as one of the pieces in the exhibition, *Dark Passage*, plays homage to Rodin's likewise life-size *Burghers of Calais*, largely known through bronze casts.

Plastics textiles and sheeting have in particular increased the scale of sculpture. Christo and Jean-Claude used industrial quantities in their 'wrapped' sculptures: *Running Fence* (1972–1976) used 200,000 square metres of woven nylon; *The Pont Neuf Wrapped* 1(1975–1985), 41,800 square metres of woven polyamide; and *Surrounded Islands* (1980–1983, which involved the encirclement of eleven islands in Miami's Biscayne Bay), 603,850 square metres of floating pink polypropylene (Christo and Jean-Claude n.d.). Anish Kapoor's massive sculpture *Marsyas* displayed in Tate Modern's Turbine Hall in 2003 is made up of a single span of plasticised PVC membrane, 135 metres in length with a surface area of 3500 metres squared (Tate n.d.-c). The functional qualities of these plastics materials played an essential enabling part in the actualisation of these sculptors' concepts.

An entirely new form of sculpture that plastics have made possible is the inflatable, exploiting not only plastics' potential for large-scale works but also for relatively cheap manufacture. An example is Paul McCarthy's work *Blockhead*, featuring ambiguous yet explicit forms in garish socio-political critiques sometimes eliciting public and critical distain in view of their appearance, subjectivity and public siting (Tate n.d.-d; Jones 2014). The art critic Martin Gayford said of them: 'Something disturbing has happened outside Tate Modern. Two huge inflatable figures have appeared, one black, one pink; both strange, not to say repulsive... The associations they immediately bring to mind are all low, down-market ones: junk food, theme parks, kitsch toys. And the subliminal associations are nastier yet—sexual and scatological' (Gayford 2003).

The use of found objects is widely embedded in sculptural practice. Plastics objects have added a new dimension partly influenced by how plastics are perceived, as demonstrated by Gayford's response to McCarthy's inflatables. Tony Cragg (n.d.) used the developing negative attitude to plastics in the 1980s to comment on consumerism through his colourful arrangements of found plastics objects (also referred to in Chap. 4), arranged variously in monochrome, figurative or abstract patterns, both on the floor and wall mounted (Cragg n.d.). More recently, Jessica Stockholder has specialised in using reclaimed plastics objects to comment on our wasteful society as in her mobile *Set Eyes On* (2014) (Mitchell-Innes and Nash n.d.). Indeed a large area of sculptural practice has built up around recycled plastics. Examples are the work of Khalil Chishtee (Recyclart n.d.) and Grace Ebert (Murmure 2020) both of whom create sculpture from plastic bags and Aurora Robson (Robson n.d.) who works with a wider selection of recycled plastics debris.

Related to the use of found objects is Jeff Koons' fascination with the re-contextualisation within gallery spaces of brand new consumer products, made, as consumer products tend to be, out of plastics. An example

is his 1980 series of sculptures comprised of wall-mounted vacuum cleaners displayed in beautifully fabricated and lit acrylic boxes (Tate n.d.-e).
The 2016 exhibition of Damien Hirst's collection of Koons's work at the
London Newport Street Gallery (run by Hirst himself) was of special significance in developing dialogue around plastics as a sculptural medium.
The exhibition displayed a broad selection of Koons's sculptures: his readymades, inflatables, large shiny steel forms that look like plastics inflatables,
and an apparently massive pile of multi-coloured plastic Play-Doh (in fact
made of spray-painted cast aluminium) (Newport Street Gallery 2016).
The juxtaposition of these works, the materials of which they are made,
and the way in which their appearance belie the materials, demonstrated a
variety of ways in which a sculptor's chosen materials, and plastics in particular, can directly, and indirectly, contribute to a work's discourse.

Translucency in Plastics Sculpture

One of the reasons plastics have been used by sculptors, of particular significance to the author's work, is their potential translucence. Potential in
so far as the maintenance of translucence in worked plastics is a quality that
has to be fought for, or at least meticulously retained, as the working of it
can render it opaque.

Translucence is not unique to plastics; translucent materials have been
used throughout history. Piero de' Medici is said to have commissioned
the youthful Michelangelo to fashion a statue of snow although to what
extent translucence was sought after is uncertain (Rich 1973, 335). There
is, however, no doubt that the architectural exploration of translucence in
glass reached a peak in the sublime stained glass of Medieval cathedrals like
Chârtres. Later, treatises such as Neri's *The Art of Glass* became available
(translated into English in 1662) (Neri 2003) feeding off existing exem-
plars and inspiring future applications of glass as a creative medium extend-
ing to the twentieth and twenty-first centuries. Examples of the use of
glass in sculpture are Sidney Waugh's figurative *Atlantica* (1938–1939)
(Corning Museum of Glass n.d.); Annie Cattrell's glass lung sculpture,
Capacity (2007) (Carel 2017); and Tony Cragg's *Visible Men* (2009)
(Glasstress 2019). Both Cattrell and Cragg are sculptors also with an affin-
ity for plastics. Additionally, chemical reactions have been used to explore
translucent effects such as Tadao Ando's *Water* (2011) (Wallace 2011),
Andy Goldworthy's *Ice Works* (Goldsworthy 1990), Olafur Eliasson's *Fog*

Assembly (2016) (Eliasson n.d.), Hans Haacke's *Condensation* (1965) (MACBA) and Matt Dabrowski's *Smoke* (Mansfield 2011).

The Russian constructivist Naum Gabo was one of the first sculptors recorded as working in plastics, and interestingly from the outset he chose translucent materials, initially working with cellulose acetate and, when it became available, acrylic. Complicating matters of precedence is the fact that often a sculpture would be made using cellulose or glass which may later, or in a subsequent version of the same sculpture, be replaced with polymethyl methacrylate (PMMA), known by the trade names Plexiglas in the United States and Perspex in the United Kingdom and often called acrylic, the term used throughout this chapter (A&C Plastics n.d.). Gabo was quick to appreciate the planar qualities of plastics materials in such works as *Monument for a Physics Laboratory*, *Tower*, and *Column* dating from between about 1922 to 1924, and soon also realised their potential for curved surfaces as in *Spiral Theme* of 1941 (Hammer and Lodder 2000, 101–125, 269–313).

An impetus to the take up of specifically transparent plastics as a creative medium was a sculpture exhibition at the 1939 New York World's Fair. The exhibition was a collaboration between the Museum of Modern Art (MOMA) and the industrial plastics corporation, Rohm and Hass, which had recently developed acrylic, the only plastics material that remains truly transparent when made into objects more than a few inches thick. The call for competition entries stated that it was hoped 'that a new sculptural technique may be developed which will express the unique properties of the plastic material' (O'Steen 2016). Alexander Calder's winning submission (see All My Eyes 2011) among 250 entries, deftly exploited the new material's transparent properties and curvilinear potential in an essentially constructivist composition.

Thereafter, minimalist sculptors were in particular drawn to acrylic for its clean surfaces and modern appearance. Louise Nevelson explored the material in her transparent modular box structures such as *Transparent Sculpture IV* of 1968 which exploited the play of light within their crystalline forms (The Art Story n.d.). Later Donald Judd contrasted the materiality of acrylic with that of aluminium in works like *Untitled* (1980) consisting of vertically suspended stacks. It offers 'viewers two conflicting experiences—opaque intrusive forms from the side, and obscure depths of space from the front' (See 2019).

Bruce Beasley was one of the first sculptors to explore the potential of casting transparent acrylic. His abstract *Apolymon* (1968–1970) was on an unusually grand scale for the material, measuring over two metres high and four metres wide. In order to create this piece, he pioneered the use in artistic circles of an autoclave, a large heated pressure vessel, which enables large blocks of plastics to cure successfully when otherwise they would crack during the process (Beasley 2014; Fernàndez-Villa et al. 2012). Others who have worked with cast acrylic include the early work of Hesketh whose works D. Cadenza (n.d) in amber coloured plastic and Form in Blue (n.d) (specific plastic not specified) were stylized abstract figurative forms (Rich 1973).

Translucent sculptures have also been made in other plastics materials. Glass reinforced polyester resin was a favourite material of Eva Hesse as in the 19 bucket-like forms comprising *Repetition 19. III* executed in 1968 (MoMA). Rachel Whiteread is renowned for her cast work in a range of media, especially concrete. The use of polyester to cast the voids under a 100 different chair in *Untitled (One Hundred Spaces)* of 1995 introduced translucent colour to her repertoire (Tate n.d.-f).

In respect to the pecuniary and market value of plastics in sculpture, an art business journal article of 2003 suggested that 'acrylic has become sculpture's material of the moment' on account of its new colors [*sic*], finishes and forms.' The article implies that the bright and varied colours by then available in acrylic had an artistic and evidently a commercial value that may in turn have inclined galleries to select sculptors using plastics and encouraged sculptors to work with them (Hagen 2003).

CAM/CAD Processes

The sculpture discussed so far has been made using traditional sculptural techniques such as casting, moulding, carving, hand-fabrication, and arranging found objects. Even inflatables involve established industrial techniques. Of particular relevance to the author's work, however, are sculptures that have employed design and manufacturing methods dependent on the use of computer software and computer-controlled machinery to automate the design and manufacturing process.

In the author's knowledge, Alfred Duca in collaboration with MIT was, in about 1968, the first artist to use a computer programme to create a sculpture. It instructed gas-cutting equipment to cut zig-zag profiled steel rings creating a three ton, seven-foot-wide creation entitled *Steel Spheroid* (Facsimile Magazine 2008).

Computer Numerically Controlled (CNC) machinery was demonstrated in 1952 at MIT (McDonough and Susskind 1952, 133). However, it is thought that the first sculptor to use a computer programme in conjunction with an engineering milling machine was Charles Csuri who in 1968 collaborated with an engineer colleague at Ohio State University to produce a sculpture in wood named after the process of its creation: *Numeric Milling* (The Ohio State University n.d.).

Again, as far as the author is aware, Robert Mallary was the first to develop his own dedicated computer programme for assisting in the production of sculpture. However, his programme (a Fortran CAD programme called Tran 2) was written to create accurate templates, which were subsequently laid out on wood-based sheet material to be cut out by hand or band saw (Mayor n.d.). Similarly, José Luis Alexanco produced his MOUVNT series using computer-generated template profiles, however, in this case, although the sheets were again worked by hand, the material used was a plastics resin resulting in a translucent sculpture (Aparicio n.d.).

Masaki Fujihata has a particular interest in exploring new technologies in association with plastics. It is thought that in 1987 he was, with his sculpture *Geometric Love*, the first to machine Ureol foam for sculptural effect using CNC (Fujihata n.d.-a). Ureol boards can be bonded together to make larger work if required, and is easy to fill and spray finish in any colour if desired. More recently, Daniel Widrig has created sculptures from computer models of cloud formation in polystyrene foam shaped by CNC machinery (Crafts Council n.d.), also a material available in sheet and block.

In 1989, Fujihata was also among the first to see the sculptural potential of stereo-lithography, a 3D-printing method invented a few years earlier. It is a process that enables the rendering of a 3D CAD model in real three-dimensional space. He used it to realise his Arpesque group of forms entitled *Forbidden Fruits* executed in translucent plastics resin (Fujihata n.d.-b). Many sculptors have since explored plastics materials delivered via 3D-printing processes to create forms not otherwise achievable. Karin Sander's miniature *3D Body scans* (1997) of human figures provide an unusual example. She explained her process thus: 'People are laser-scanned using a body scanner that employs a 3D photographic process... Their data is then sent to an extruder, which recreates their body shape slice-by-slice in plastic... [creating] an exact reproduction of the person in question... The figure is produced entirely by mechanical means...' (Sander n.d.). By contrast, Keith Brown's 3D-printed topological sculptures create a cyber environment located in an area that exists beyond the imagination and everyday experience. He has said of his work, 'The use of computing technologies is an essential aspect of my creative practice and is indispensable to the conception, content, and quality of the artwork' (Manchester School of Art n.d.).

These few examples make it clear that computer assisted design and manufacture in plastics has opened up a new dimension in sculptural practice that is continuing to develop.

THE AUTHOR'S PLASTICS' PRACTICE

The author's sculptural practice presents a confluence of formalist and conceptual ideas. The formal influences draw on a range of ideas of long pedigree. They include early classical notions of proportion deriving from Pythagorian philosophy, Euclidean geometry, Platonic ideas concerning light and *formes* and the Aristotelian notion of entelechy (the basic existential ontological state of matter) (Sachs 1995, 24, 245; Sachs 2002, 77) as well as Non-Euclidean geometry derived from the mathematical fields of calculus and topology. Conceptual influences relate to how employment of the transparent qualities of cast-acrylic aligns itself with Scholastic, particularly Thomist, thought. For Aquinas, conceptions of beauty were to be found in the qualities of '*integritas, consonantia,* and *claritas*' [integrity or wholeness, proportion and radiance or clarity]. A related influence is the anagogical notion of art leading the viewer to the transcendency delineated by Scholastics such as Abbot Sugar and Hugh of St. Victor (Taylor-Coolman 2010, 44).

A specific influence on the author has been the Anglican hymn *Teach Me My God and King* by George Herbert (1593–1633), the words of which specifically and poetically align the notion of clarity redolent in Thomian thought, with the anagogical thought of Abbott Sugar and Hugh of St. Victor (Taylor-Coolman 2010, 44). These two thoughts are drawn together with the analogy of the translucent materiality of glass:

A man that looks on glass,
on it may stay his eye;
or if he pleaseth, through it pass,
and then the heaven espy.

In his commentary on this hymn, Middleton proposes that 'The paradox of mystery and revelation, transcendent and imminent, is beyond
comprehension yet it is revealed through those images which can contain
the truth of both states at once' (Middleton n.d.). Such Thomian qualities
of clarity have been assigned to certain paintings or frescos, for example,
those by the fifteenth-century Italian monk, Fra Angelico, as well as to
stained glass windows, as previously mentioned. This was presumably in a
metaphorical or spiritual sense as a function of qualities of light implied by
the use of, and relative saturation of, certain paint colours; and the generally sympathetic treatment of certain, predominantly biblical subjects. It is
argued here that plastics (and other translucent materials) permit these
two states more readily than painted canvas and though unavailable at the
time, cast acrylic (like glass) is a material well suited to express Thomian
aesthetics.

The author has explored in particular the semantic and structural
potential of cast acrylic in pursuit of the sculptural representation of transcendency. It has technical properties that enable it to be machined to a
high degree of accuracy, abraded and subsequently polished to achieve a
highly translucent quality. It also has a semantic value insofar as its potential translucence allows the viewer to see both its surface, its interior and
also to see through it entirely as the words of Herbert's hymn observed
(n.d.). These visual properties destabilise the eyes' response to it and create a perceptual dissonance with, on the one hand, a sense of solidity, presence, immanence and permanence yet on the other a sense of mutability
and, the argument is advanced, transcendence. Furthermore, by altering
the surface texture, the level of translucence can be controlled and,
thereby, light can be introduced into the internal volume of the sculpture
as well as reflected off the surface. Moreover, with greater translucence,
forms and movement can be detected on, through and beyond the object.
All these properties are significant in relation to the author's work.

However, the material presents challenges from a manufacturing point
of view. In view of its relative intractability to conventional sculptural
tools, the author, like others as discussed above, has employed CAD/
CAM procedures as an agentive means to work pre-cast acrylic reductively,
from a solid block. His preference for reductive working rather than casting (creating form by the setting of poured liquid) is because the latter is
necessarily a secondary process, requiring an initial mould and thereby can
lack the crispness of machined form. Thus, it is argued that the relatively
recent availability of digital manufacturing methodologies has enabled the

exploration of precise geometries in cast acrylic expanding the realisation of formalist conceptions.

The author has appreciated the value of plastics as a sculptural material since undertaking cast polyester work dating back to 2003. However, the work focused on here relates to cast acrylic in conjunction with CAD/CAM methodologies in relation to three sculptures made from 2008 to 2010.

The first of the group, *Sine Bowl* of 2008 (Fig. 3.1) represents a formal attempt to subvert the traditional notions of a bowl by introducing geometric distortions that challenge the conventional understanding of a shallow vessel form.

The piece draws on the neo-classical formalist tradition of employing the kind of mathematical relationships evident in classical architecture such as the Parthenon, evoking sculptors such as Max Bill who sought 'to develop an art largely on the basis of mathematical thinking' (Smith 1994). This mathematical underpinning has been actualised by direct access to the mathematical algorithms of CAD software and computer-mediated numerical control. It provides thus, also, an investigation into the

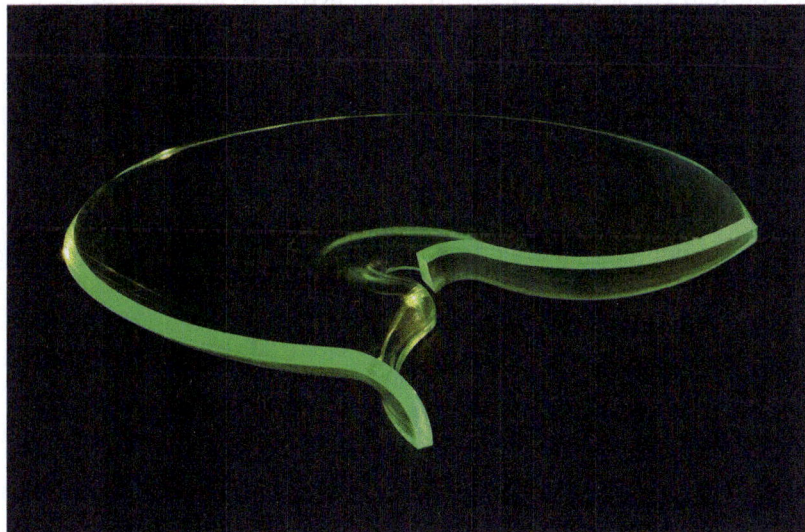

Fig. 3.1 *Sine Bowl*, 2008, cast acrylic; diameter: 400 mm; height: 70 mm. (Photo: Richard Hooper)

potential of CNC technology to create a form in three axes following earlier precedents discussed above.

The material selected is also significant. The green translucent cast acrylic symbolises new growth, adding to the dynamic aspect of the form. The plastics' translucency amplifies the visual weight of the linear edge detail and the optical effect of the material serves to concentrate the light at this point. The translucence furthermore has the effect of reducing the solidity of the form and gives it a somewhat ethereal archetypal quality.

The sculpture, *He Is with You* (Fig. 3.2), made the following year, was commissioned by churchwardens as a present for a retiring vicar and included a requirement to use a piece of oak left over from the reordering of the church. The subject matter was inspired by the biblical story of Christ calming the storm on the Sea of Galilee. His fearful disciples woke him to alert him to the danger of being overwhelmed by the waves. Christ then, having rebuked them for their lack of faith, stood and prayed for the storm to abate (Matthew 8: 23–27).

The form of the boat is derived from a putatively contemporaneous boat depicted by the surrealist Salvador Dali in his painting *The Christ of St John of the Cross* (1951). The cast acrylic used to represent the water was intended to give a vibrant translucent counterpart to the textured surface of the boat's wooden form. From a semantic point of view, it enabled the wave to be rendered transparent and analogous to the omniscience of God's knowledge of all life's circumstances. At the same time the translucent plastics conveys a sense of the visceral liquidity, otherness and transparency of the water. From a formal point of view, a variety of curvatures were explored, including the wave element, before a relatively simple sine wave with counter-posed angles at the extremities was chosen to complement the more symmetrical boat form. The inclined angle of the boat gives the appearance of the boat in the heavy swell and enables a clearer view of the hull geometry. From a technical point of view, the piece made use of the CAD/CAM capability of Boolean difference whereby the geometry of a positive form may be subtracted from another form allowing the two mating surfaces to be made and fitted together (like a horse chestnut or avocado nut in its protective casing). This capability of subtraction enables a variety of sculptural possibilities. The form of both boat and its interaction with the wave demanded complex geometrical manipulation for which CAD/CAM, with its mathematical basis, proved ideal.

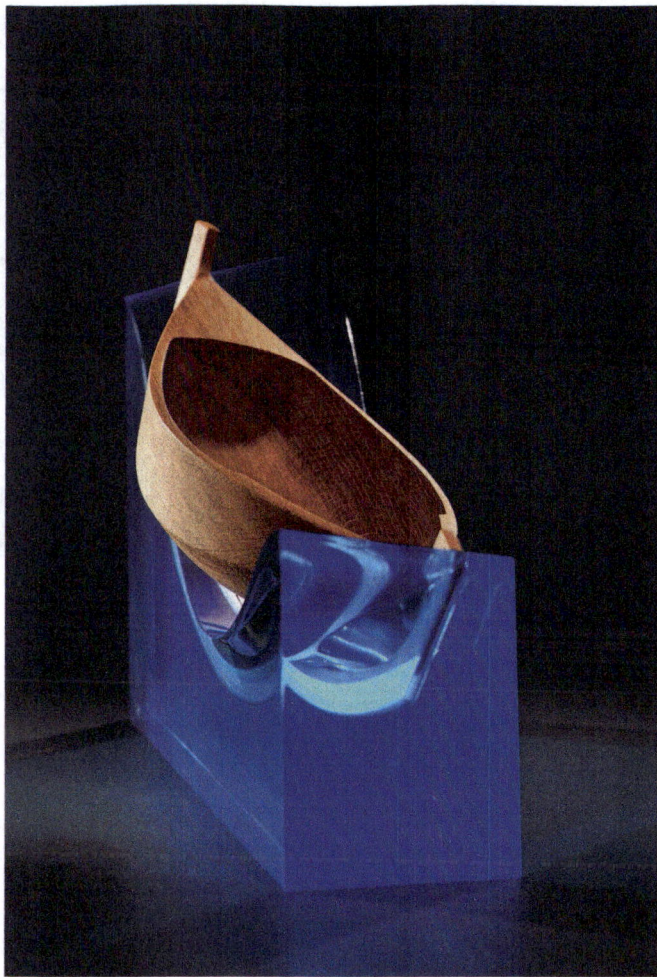

Fig. 3.2 *He Is with You*, 2009, oak and cast acrylic; length: 300 mm, height: 70 mm, width: 100 mm. (Photo: Richard Hooper)

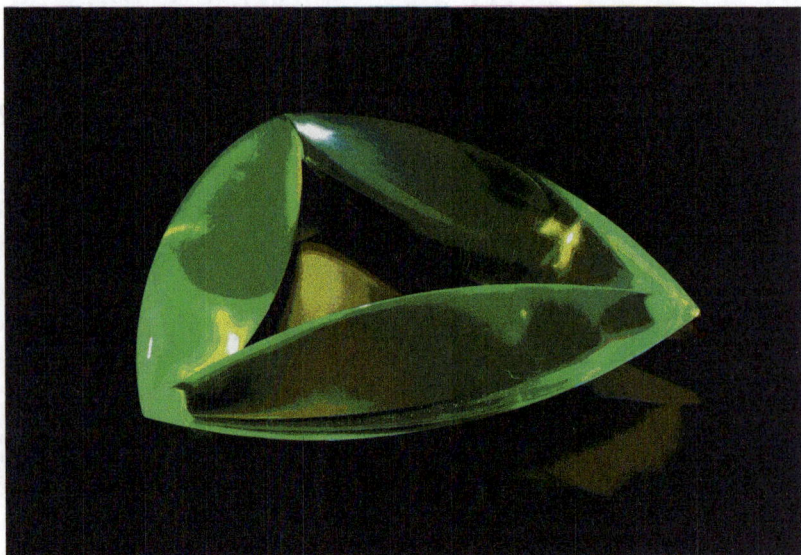

Fig. 3.3 *Trinity*, 2010, cast acrylic; length: 300 mm, height: 70 mm, width: 300 mm. (Photo: Richard Hooper)

The sculpture, *Trinity* (Fig. 3.3) dating from 2010, exploits especially the translucence of cast acrylic with a view to evoking Thomist qualities of clarity and luminosity.

It has a tripartite form of repeating elements representing the Father, Son and Holy Spirit of Christian theology in an overtly formalist abstract, quasi-futuristic, idiom. In respect to the use of analogous formal archetypes as an expository device, Saint Patrick is said to have used the three-leaved Irish shamrock to explain the nature of the Trinity in his early British preaching (Van Sloun 2011). The bright green colour, symbolic of growth and new birth, gives a freshness and organic feel to the piece with its free-flowing internal volumes and liminally ambiguous relationship with external space.

The formal composition of the work is based on three conjoined variations of the vesica piscis (literally 'fish bladder' in Latin) lens shape. Where the classical vesica piscis has two equal circles such that the centre of each circle lies on the circumference of the other, those in this piece are spaced so that the height of the major axis of the modified lens is equal to half the height of the conventional vesica piscis (giving a slimmer lens shape). This shape is then bisected longitudinally and the resulting half rotated through 360 degrees (in the CAD program) to create a solid form. The forms are then obliquely truncated at both ends by two mirrored planes set at 30 degrees to the major axis and intersecting at a point halfway along the radius (set at 90 degrees to the major axis) of a circle circumscribing the modified lens shape. The truncated faces of the three forms are then mated to construct the final three-dimensional abstract form. Only the combination of cast acrylic manipulated by CAD and CAM made this possible.

The sculpture (Fig. 3.4) was ultimately used as a component of an installation in which the translucent acrylic sculpture was placed in the snow in a churchyard, with the church in the background. The whole installation sought to convey the immanence of the triune God descended to the earthly material realm; rendered in a quasi-science fiction 'landing' composition. Again, the translucence of cast-acrylic provided an ideal material and enabled both precise geometric fidelity to the CAD model and a vibrant translucency with which to explore the sculptural concept and embody its meaning.

CONCLUSION

It has been shown that sculptors have used plastics for the full gamut of sculptural practice: figurative, abstract and conceptual work. Plastics have enabled realism to be taken to a new level, set sculptors' imaginations free in terms of the potential scale of their work, introduced the inflatable, and expanded conceptual practice in relation to found objects and readymades through values attached to them. They have also introduced translucency as a standard constituent of sculptural vocabulary. In addition, it has been shown that a combination of plastics, computer modelling and computer manufacturing, whether in terms of milling machines or 3D-printing, has enabled a new approach leading to the creation of form of which it was formerly difficult to conceive and impossible to achieve.

The chapter has also demonstrated importantly in respect of the author's work the potential of plastics as a machinable material of particular utility where CAD/CAM methodologies are employed, especially when incorporating Euclidean or Non-Euclidian geometries. The application of these methodologies to translucent cast acrylic has expanded the range of formal geometric possibilities allowing vistas in and through the

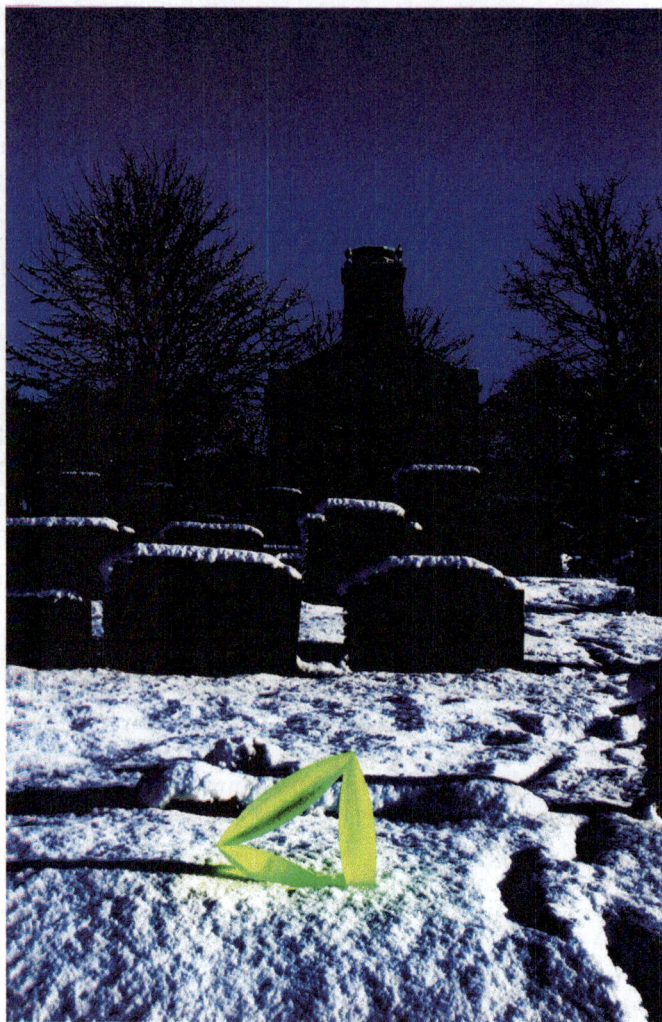

Fig. 3.4 *Trinity*, as a Photoshop-enhanced image of a site-specific installation, 2011. (Photo: Richard Hooper)

material consonant with Thomian qualities and as expressed in Herbert's hymnal vision.

It is postulated that plastics (subject to their acceptability on environmental grounds) will continue to play a significant role in sculpture as a result of their technical properties and semantic potential. Furthermore, where there is need for clarity, translucency or geometric fidelity, or where transparency is important as a vehicle to convey transcendent themes, cast-acrylic will remain a medium of exceptional sculptural value.

BIBLIOGRAPHY

A&C Plastics, Inc. n.d. Plexiglas®: Acrylic Sheeting Facts. Accessed 28 August 2019. https://www.acplasticsinc.com/informationcenter/r/plexiglass-vs-Plexiglas.

All My Eyes. 2011. Modern Sculpture Made Clear. Accessed 27 August 2019. http://allmyeyes.blogspot.co.uk/2011/07/modern-sculpture-madeclear.html.

Aparicio, Carmen Fernàndex. n.d. Mouvnt Sculpture. Accessed 23 March 2019. https://www.museoreinasofia.es/en/collection/artwork/escultura-mouvnt-mouvnt-sculpture-1.

Beasley, Bruce. 2014. Apolymon. Accessed 3 June 2019. http://brucebeasley.com/2014/11/03/apolymon-2/.

Carel, Havi. 2017. The Uses and Abuses of Air. Accessed 3 March 2019. https://lifeofbreath.org/2017/06/the-uses-and-abuses-of-air-2/.

Christo and Jeanne-Claude. n.d. Accessed 23 August 2019. https://christojeanneclaude.net/.

Corning Museum of Glass. n.d. Collection. Atlantica. Accessed 27 August 2019. https://www.cmog.org/artwork/atlantica.

Crafts Council. n.d. Makers > Daniel Widrig. Accessed 13 June 2019. https://www.labcraft.org.uk/makers/about/daniel-widrig.

Cragg, Tony. n.d. Self-portrait on a Chair. Accessed 23 March 2019. http://www.tony-cragg.com/?/sculptures/1980-1984/930.

Eliasson, Olaf. n.d. Fog Assembly. Accessed 23 August 2019. https://olafureliasson.net/archive/artwork/WEK110139/fog-assembly#slideshow.

Facsimile Magazine. 2:1. 2008. Automated Carving in Steel. Accessed 23 August 2019. http://facsimilemagazine.com/2008/01/index.html.

Fernàndez-Villa, Silvia Garcì et al. 2012. Industrial Development of Plastics and 20th-Century Art: New Synergies. Accessed 12 December 2019. https://www.researchgate.net/publication/322581451_Industrial_development_of_plastics_and_20th-century_art_new_synergies#pfl.

Fujihata, Masaki. n.d.-a Geometric Love. Accessed 1 July 2019. https://www.youtube.com/watch?v=Lw7yc6mIRqA.

———. n.d.-b. Forbidden Fruits. Accessed 1 July 2019. https://www.youtube.com/watch?v=yygElySiA-.

Gayford, Martin. 2003. *The Telegraph*, May 29. Accessed 15 March 2020. https://www.telegraph.co.uk/culture/art/3595535/Whats-black-and-pink-and-blows-up.html.

Glasstress. Tony Cragg. n.d. Accessed 16 August 2020. http://glasstress.org/my-product/tony-cragg/.

Goldsworthy, Andy. 1990. *A Collaboration with Nature*. London: Abrams.

Hagen, D. 2003. Clear Outlook for Acrylic Sculpture in Art. Accessed 27 June 2019. https://www.questia.com/magazine/1G1-109905472/clear-outlook-for-acrylic-sculptures-with-technology.

Hammer, M., and C. Lodder. 2000. *Constructing Modernity; the Art and Career of Naum Gabo*. New Haven: Yale University Press.

Herbert, George. n.d. Teach Me My God and King. Accessed 4 April 2019. https://danutm.wordpress.com/2015/03/17/teach-me-my-god-and-king/.

Jones, Jonathan. 2014. *The Guardian*, October 20. Accessed 20 March 2019. https://www.theguardian.com/artanddesign/jonathanjonesblog/2014/oct/20/paul-mccarthy-paris-tree-sculpture-butt-plug-controversy.

Karmel, Pepe. 2017. Robert Morris: Boustrophedons. Accessed 4 April 2019. https://www.castelligallery.com/attachment/en/5ccc8899a5aa2c2f30864af2/Publication/5d126c9e72a72c2116f961c4.

Lillian Goldman Law Library. 2008. Military-industrial Complex Speech, Dwight D. Eisenhower. Accessed 23 August 2019. http://avalon.law.yale.edu/20th_century/eisenhower001.asp.

Lucie-Smith, Edward. 1987. *Sculpture since 1945*. London: Phaidon Press.

MACBA. n.d. Hans Hacke. Condensation Cube. Accessed 3 December 2017. https://www.macba.cat/en/condensation-cube-1523.

Manchester School of Art. n.d. Keith Brown. Sculptor Profile. Accessed 4 August 2019. https://www.art.mmu.ac.uk/profile/kbrown.

Mansfield, Guy. 2011. Unstable Arrangement Smoke Sculpture by Matt Dabrowski. Accessed 2 December 2017. https://www.youtube.com/watch?v=rsQjRGG4Zso.

McDonough, J., and A. Susskind. 1952. Numerically Controlled Milling Machine. *Managing Requirements Knowledge, International Workshop on*, New York City. Accessed 23 June 2019. https://doi.ieeecomputersociety.org/10.1109/AFIPS.1952.14.

Middleton, Arthur. n.d. Faith of our Fathers. Extraordinary in the Ordinary. Accessed 4 March 2019. http://trushare.com/91DEC02/DE02FAFA.htm.

Mitchell-Innes and Nash. n.d. Jessica Stockholder. Accessed 4 December 2019. https://www.miandn.com/artists/jessica-stockholder?view=slider#33.

Murmure. 2020. Black Trash Bags Take Control of Animal and Human Life in 'Garb-age'. Accessed 10 May 2020. https://www.thisiscolossal.com/2020/03/murmure-black-garbage-bags/.

National Galleries Scotland. n.d. Art and Artists: Ron Mueck. Accessed 2 July 2019. https://www.nationalgalleries.org/art-and-artists/94045/girl.

Neri, Antonio. 2003. *The Art of Glass*. Hubbardston, MA: Heiden and Engle.

Newport Street Gallery. 2016. Jeff Koons: Now. Accessed 23 January 2020. https://www.newportstreetgallery.com/exhibition/jeff-koons-now/.

O'Steen, D. 2016. Presenting Plexiglas: Alexander Calder, Gilbert Rohde, and the Museum of Modern Art at the 1939 New York World's Fair. Accessed 2 September 2019. https://www.youtube.com/watch?v=pAShWaynw-A.

Read, H. 1964. *A Concise History of Modern Sculpture*. London: Thames and Hudson.

Recyclart. n.d. Discarded Plastic Bags. Sculptures by Kkalil Chistree. Accessed 9 June 2019. https://www.recyclart.org/discarded-plastic-bags-sculptures-by-Khalil-chishtee/.

Rich, C. 1973. *The Materials and Methods of Sculpture*. Oxford: Oxford University Press.

Robson, Aurora. n.d. Sculpture. Accessed 7 June 2019. http://www.aurorarobson.com/sculpture/.

Sachs, Joe. 1995. *Aristotle's Physics: a Guided Study*. Chicago: Rutgers University Press.

———, ed. 2002. *Aristotle: Nicomachean Ethics*. Indianapolis and Cambridge: Focus Publishing.

Sander, Karin. n.d. 3D Body Scans. Accessed 8 August 2019. https://www.karin-sander.de/en/work/3d-bodyscan.

See, Grace Ignacia. 2019. Famous Minimalist Art That Defined the Genre. Accessed 10 May 2020. https://theartling.com/en/artzine/famous-minimalist-art/.

Serpentine Gallery. 1978. Henry Moore at the Serpentine: 80th Birthday Exhibition of Recent Carving and Bronzes. Accessed 9 June 2019. https://www.serpentinegalleries.org/whats-on/henry-moore-serpentine-80th-birthday-exhibition-recent-carvings-and-bronzes/.

Smith, Roberta. 1994. Max Bill, 85, Painter Sculptor and Architect in Austere Style. *The New York Times*, December 14. Accessed 23 January 2020. https://www.nytimes.com/1994/12/14/obituaries/max-bill-85-painter-sculptor-and-architect-in-austere-style.html.

Tate. n.d.-a. Naum Gabo. Spiral Theme. Accessed 2 February 2019. https://www.tate.org.uk/art/artworks/gabo-spiral-theme-t00190.

———. n.d.-b. Art and Artists: William Tucker. Accessed 2 February 2019. https://www.tate.org.uk/art/artworks/tucker-anabasis-i-t01200.

———. n.d.-c. Exhibitions. Unilever Series: Anish Kapoor; Marsyas. Accessed 24 June 2019. http://www.tate.org.uk/whats-on/tate-modern/exhibition/unilever-series/unilever-series-anish-kapoor-marsyas.

———. n.d.-d. Exhibitions. Paul McCarthy. Accessed 20 March 2019. http://www.tate.org.uk/whats-on/tate-modern/exhibition/paul-mccarthy.

———. n.d.-e. Jeff Koons. New Hoover Convertibles, Green, Red, Brown, New Shelton Wet/Dry 10 Gallon Displaced Doubledecker. Accessed 2 February 2019. https://www.tate.org.uk/art/artworks/koons-new-hoover-convertibles-green-red-brown-new-shelton-wet-dry-10-gallon-displaced-ar00077.

———. n.d.-f. Art and Artists: Rachel Whiteread. Accessed 2 February 2019. https://www.tate.org.uk/whats-on/tate-britain/exhibition/rachel-whiteread.

Taylor-Coolman, Boyd. 2010. *The Theology of Hugh of St. Victor.* Cambridge: Cambridge University Press.

The Art Story. n.d. Louise Nevelson. Accessed 12 July 2019. https://www.theartstory.org/artist/nevelson-louise/artworks/#pnt_5.

The Mayor Gallery. n.d. Writing-New-Codes-Catalogue. Accessed 5 June 2019. https://www.mayorgallery.com/artists/157-robert-mallary/.

The Ohio State University. n.d. Charles A. Csuri Project. Accessed 11 July 19 2019. https://csuriproject.osu.edu/index.php/Detail/objects/769.

Van Sloun, Michael. 2011. St Patrick, the Shamrock and the Trinity. Accessed 4 June 2018. http://catholichotdish.com/the-pastors-page/saint-patrick-the-shamrock-and-the-trinity/.

Wallace, William. 2011. Tadao Ando. Water installation—Outside the Connaught Hotel London. Accessed 13 August 2019. https://londoniscool.com/tadao-ando-water-installation-%E2%80%93-outside-the-connaught-hotel-london.

Wroe, Nicholas. 2014. Allen Jones: 'I Think of Myself as a Feminist'. Accessed 5 June 2019. https://www.theguardian.com/artanddesign/2014/oct/31/allen-jones-i-think-of-myself-as-a-feminist.

CHAPTER 4

Can Plastics Be a Muse for Future Feminist Innovation?

Flora McLean

I have a fashion label called House of Flora and make hats and accessories that sell worldwide to stores and galleries. I studied Fashion Design at the Royal College of Art, and was known by my peers as 'Plastic Fantastic', and am now head of its Footwear Accessories Millinery and Eyewear (FAME) specialism. Plastics inspire me as a materials group because of the infinite possibilities of form that they can create, re-position, and mass produce as art objects or wearables. I see plastics as a malleable force for moulding the infinite possibilities of my imagination to propose a cool synthetic future aesthetic for warrior-like women.

Deborah Jaffé has written on the practical influence plastics have had on the emancipation of women over the past 150 years. For example, she introduces the idea that the bicycle, and specifically the Dunlop rubber tyre that allowed them to become popular, was one of the first ways in which women were liberated to travel alone, with privacy. Once women were free to travel without chaperones, no one had to know where to or

F. McLean (✉)
Royal College of Art, London, UK

House of Flora, London, UK
e-mail: flora.mclean@rca.ac.uk

69

why a woman was travelling, and this freedom greatly advanced the fight for women's freedom and self-actualisation (Jaffe 2009). By contrast, my subject is, rather, how through the agency of plastics women can be empowered and transformed.

Catwoman, particularly Michelle Pfeiffer's rendering in *Batman Returns*,[1] which I will come back to, is the cornerstone of my conception of plastics as an empowering and transformative materials group. My case is that plastics have not only the power to transform women but also to enable them to find their own, individual routes to power and self-actualisation, and thus to be a muse for future feminist innovation. This chapter explores this emancipatory role of plastics by referencing art, films and theories as well as my own practice and my archive of hats and accessories. It also draws on interviews with individuals who employ plastics not just as a tool or a vehicle of self-expression (as I do), but further, have and continue to ingrain it into their very personhood, using plastics not only to augment their physical selves but as Catwoman does, to define for themselves a new mode of being.

PLASTICS, A MATERIAL OF REBELLION

I originally studied fashion. This was, however, an accident. I could not decide where to specialise, and fashion was just one of my interests. I feel privileged to come from a family of architects, conceptual thinkers, fine artists, and textile designers, and therefore my work has been heavily influenced by a wide spectrum of colours, ideas and theories. During my foundation year at art school, I was put in the textile and fashion workshops but felt far more at home in the spatial design and sculpture workshops. Now I am what is technically termed a 'milliner', specialising in making hats for fashion, but prefer to be called a fashion designer who makes products.

In the early 1990s, I designed a beret dip-moulded in PVC, titled at that time as the Jelly Beret, which continues to be made today, over 20 years later, in multiples in a UK factory, the main business of which is manufacturing plungers and medical equipment. My passion for accessories is based on my belief that it is the details, which are the point in fashion. The early marketing of this particular hat was a portrait in the style of Che Guevara whose revolutionary image became a symbol of rebellion for global culture (Fig. 4.1).

Fig. 4.1 Che Guevara-
style marketing image
for the *Jelly Beret*.
Photobooth 1999.
(Photo: Flora McLean)

The beret provides an example of rebellion through fashion: wearing any hat is provocative, but wearing a smooth, futuristic hat made of plastics takes it to another level entirely.

Most fashion designers would not associate plastics materials with their métier other than as materials for linings, undergarments and lingerie or as substitutes for more expensive 'natural' materials. Nonetheless, they have been used in a number of forms: polyvinyl chloride (PVC) for rain macs, acetate for eyewear, moulded resins and acrylics for jewellery, thermoplastic supports for costumes and hats, synthetic rubberised coatings for weather-resistance and inflatable wearables for fun, safety and art pieces. A striking pioneer of innovative use of plastics in fashion was Elsa Schiaparelli. In 1934, she made what was described as a 'glass' cape from Rhodophane, a transparent plastics related to cellophane (Curtis 2014, 56) and is renowned for making a feature of cellulose acetate zippers as well as for her 'Lucite' jewellery, Lucite being a trademark for a type of acrylic. In 1938, the designer Elizabeth Hawes in her book *Fashion as Spinach* put forward the concept of pouring substances into moulds to solidify into a finished garment, one way of working with plastics (Hawes 1938, 289); and 'the impact of the revolutionary nylon, described as coal into silk, was claimed by one newspaper as 'more revolutionary than Martian attack'' (Curtis 2014, 56). Post–World War II Mary Quant capitalised on the 1960s' love affair with new materials and was the first designer to use

PVC, creating 'wet look' clothes and different styles of weather-proof boots in her footwear range, Quant Afoot (Quant 2012, 112, 120–121).

Fiorucci was a shop I used to visit as a child and teenager; an edgy 'cool' store, with fashion for young people, including accessories, which made plastics desirable, wearable and fashionable. The range of items for sale were displayed in an arty, music-filled space that felt more like a playground or an art installation than a run-of-the-mill retail spot. As Eve Babitz wrote: 'Fiorucci thrives on its perversion of form, its refusal to do things the way everybody else does them. They don't use silk everyone else does [...] In 1979, some of the Fiorucci jeans turned transparent; they were produced in see-through plastic [...] cheerful, playful galoshes in fanciful colours so happy that your mood would improve the instant you looked down' (Babitz 1980). To me it felt like a sweet shop, and exerted a huge influence over my ideas.

Another formative influence was a visit to the Anthony D'Offay Gallery where I saw a plastics installation by Tony Cragg, created from collected, collated, and categorised found everyday plastics objects. The unifying feature was that everything was orange. I remember looking at these objects: some things I had as toys or they were household objects and seemed so familiar, but they had been given a special place carefully curated on a gallery floor. Recently, Cragg recalled this aspect of his work: 'at that period sculpture was ... about finding new materials, I was interested in manmade materials ... plastic things just seemed obvious ... they were around ... sometimes a plastic object looked like it was from another planet, it looked something very beautiful' (cited in San Francisco Museum of Modern Art 2019). Seeing plastics placed in such a setting contributed to my belief in plastics as a 'special' or higher material. The fact that they were orange may also have influenced my fascination with this colour.

As a student, I began collecting orange kitchen utensils made of plastics and began hoarding them. I still search for things at car boot sales and more recently online. I love this colour. It is described by the Pantone Colour Institute as: 'From sweet smelling peach and energizing coral to vibrant tangerine, spicy ginger and earthy terra cotta the meaning of orange is inexorably linked to the sensations of radiant energy, heat and the glowing presence of the setting sun' (Pressman 2017). For me it represents spontaneity, motivation, positivity and the future (Frieze 2000) but the utensils have to be made of plastics. The reason for this is that, as I perceive the material, it represents a space age, that is, an age in the future as, for example, underpinned inspiration in the 1960s product and furniture styling revolution. Now I actually use all of these pieces at home. Although this use of plastics is not necessarily fashion wear, my 'orange kitchen', full of my collected items, has become part of my identity, my aesthetic, almost as much as my wardrobe. Perhaps I was trying to recreate the sweet-shop-feeling of the Fiorucci shop, to have my own candy shop in my kitchen, or my experience of the Cragg installation, and that, in a way, is what fashion does too. Fashion helps you capture a feeling, a moment, a memory, which can be physically distilled upon your body (Fig. 4.2).

This use of plastics as a means of designing my home aesthetic supports the view that plastics can help form self-image and contribute to personal identity.

When I studied fashion in the early 1990s, most students were concerned with cloth and the properties of textiles. Due, in part, to my lack of skills in this area (an inability to stitch and sew straight and accurately), I had no choice but to use other materials as a counter-act as I sought other ways to build form on the body. I was often to be found in 3D workshops where I was exposed both to new materials and alternative methods of construction and production (normally reserved for product design). I already saw plastics as the material of current relevance: a clean, smooth, space-related futuristic material and a counterpoint to the more traditional materials like fabric or leather; even as a rebellion against them. Polymethyl methacrylate (PMMA), better known in the United Kingdom by its trade name, Perspex and frequently called acrylic, was often my initial starting point. I also used PVC, neoprene, and latex, as well as found plastics sourced from recycling bins and offcuts. Using hard plastics and bonded synthetic fabrics, which do not require hemming or linings, I could cut raw shapes to add structure to or drape on the body. It was a practice that encouraged both new approaches and new outcomes.

Plastics are now my chosen material. This choice represents a rejection of traditional materials and techniques in millinery practice such as natural structures, like straw, felt, sinamay, and domette. To me, these materials represent a stereotypically 'occasion-wear, dusty, haberdashery' aesthetic from which I attempt to depart as a designer, in search of 'the new'. I use techniques such as pouring resin, as suggested by Elizabeth Hawes, cutting acrylic, dip-moulding PVC and 'sonic-welding' of inflatable structures, creating my personalised design language that aids the development of my unique visual aesthetic. In my studies, I worked with Nick Crosbie of Inflate, a consultancy that specialises in inflatable structures marketed as

Fig. 4.2   Author's orange plastics' kitchen, 2017. (Photo: Flora McLean)

'made in the future' (Inflate n.d.). As stated by McLean and Silver, 'The inflated elements variously gave structure to a coat collar, hat brim or the hoop of a skirt'. As architects, it is their opinion that 'The inventiveness of fashion designers in the use and development of new materials and fabricating processes, coupled with the fast turnaround of the fashion calendar, provides a great incubator of ideas that more architects would do well to embrace' (McLean and Silver 2015, 144–145). The resulting silhouettes were seen as 'different' in the fashion studio and a rejection of the more standard materials and production methods for making garments and/or accessories. Rebellion is at the heart of my practice and has a huge role to play within feminist revolution and innovation. This rejection and rebellion and going against the grain of traditional fashion practice are now what I am known for in the industry. I believe this rebellion can start before a garment has even been designed. Rebellion can be in your choice of material and my choice is frequently plastics in one form or another.

You need to be fearless in order to come up with new ideas. In the words of Mary Quant: 'The breaking of traditional rules is always exciting. Rules are made to be broken. When you break a rule you automatically arrive at something different' (Quant 2012, 137). Much of the work that I do and the way that I think are seen as controversial and cavalier. With activists participating in the global movement, Extinction Rebellion, all around me, working with plastics is part of the controversy of my work and design practice. Part of the rebellious character of plastics for me is their anti-naturalistic reputation and it is, I believe, this status as modern, anti-traditional materials that gives them the agency as transformative materials. The paradox of this reputation is the alternative view of plastics as harmful materials; related to this is the misconception that plastics are not made from 'natural materials', and as such are not perceived in the same way as woods, metals and glass. This is a subject that is returned to in Chap. 13.

The rebellious case that plastics are actually natural materials has been cogently made by the conceptual artist, Neil Cummings, who, for a while, washed and put on a shelf an example of every plastic bottle that passed through his house:

> Plastics are lodged in our consumer consciousness as completely synthetic, when in fact the materials are derived—as with everything else in our material culture—from natural sources, principally oil, natural gas, cola and salt. The origin of this confusion, and the reason we can identify a radical shift in

the history of technology is the depth of our technological intervention. As the material Bakelite and its derivatives multiplied, a German organic chemist Herman Staudinger invented a group of both natural and synthetic substances we now refer to as polymers. (Cummings 2004, 323)

Nevertheless, plastics have a tendency to be perceived as the Frankenstein of materials just because of 'the depth of our technological intervention' in that they are chemically created rather than mined or sourced directly from the earth. However, even now, when the environmental ethical implications of using plastics are very much to the fore, many of my students are trying to make their own plastics using 'natural' materials such as algae and sweet potatoes, a testimony to the value of plastics as a material in design. Their aim is to create a material that looks just like our established image of plastics: shiny and new, and to perform in the same way, except that they are looking for materials that biodegrade at a low heat and have thus low or even nil environmental impact. One student, Søs Hejselbæk, is 3D-printing modular units through an algorithm based on the flow of rivers that click together to resemble female forms. I believe the amazing variety of ways in which plastics can be conjured up combined with their infinite versatility: their textures, hues, weights and the forms they can make, imbues them with magic. In the words of the much quoted, with reference to plastics, Roland Barthes, 'in essence, the stuff of alchemy' (Barthes 2000, 97).

Feminist Innovation through Plastics

I was always influenced by my mother's style. She is more Mary Quant's era. She and I would pore over *Vogue*, trying to find a way to recreate the Claude Montana[2] plastics macs we saw on the pages with the materials we had to hand. I remember walking with my mother on the King's Road, when we noticed a glossy electric blue coat in a window. I now know it was designed by Antony Price, best known for his work for performers like the Rolling Stones, David Bowie, Duran Duran, and especially Roxy Music, and admired for his intricate and detailed structures. In the words of the journalist Chrissy Iley, 'What I have always found interesting about Price's work is the way his cut can feminise men and yet make them look more masculine. And his tailoring for women can be so masculine it's feminine' (Iley 2008). When my mother tried it on, the construction of the coat transformed her figure. The in-built corsetry gave her a sculpted shape I have never forgotten. It impressed on me fashion's potential to change how people are perceived.

I have an alter ego as a disc jockey. When I perform, I play only vinyl, always wearing a hat made of plastics. In this respect, I am influenced by Miss Poly Styrene, the punk singer from X-Ray Spex. She evoked a strong image of female rebellion, expressed through both her name and her clothes: in most images she is wearing a PVC dress and crash hat (unusual in the 1970s) (Poly Styrene—X Ray Spex n.d.). DJing is a practice dominated by men so when I DJ, I like to express my style via a hat and feel like I am controlling the room and the mood with the sound and image. The hat becomes my identity, a symbol of my rebellion and a contributor to my release from the patriarchal idyll. It also epitomises what I like about plastics: forms that are not what they look like, apparently hard but actually soft and vice versa, the trompe l'oeil effects and illusions that plastics can perform, something that is simultaneously unusual and familiar, a shiny beret in the shape of a breast misplaced on your head.

A more extreme example of empowering transformation is provided by a head piece designed by Maiko Takeda and worn by the Icelandic singer and songwriter Bjork on the cover of her 2015 album *Vulnicura* (Pitchfork n.d.). A radical, both sartorially and musically, the headpiece with its halo of candy-coloured translucent plastics spines, gives the singer a celestial radiance, as if she were a goddess from the distant future. This use of plastics is not utilitarian but atmospheric, creating form in suspended space, emitting light, and suggesting traces of movement around the wearer. The properties of plastics contribute their particular magic to the image. Is there any other material that could have achieved this so effectively?

Another fascinating persona fashioned in plastics is Pandemonia. She is a walking conceptual artwork invented in 2008 as a parody of celebrity by an anonymous London-based artist and has been vividly described by Katia Ganfield of *Vice* as 'Roy Lichtenstein's blonde caricatures … brought to life as a 7 ft Jeff Koons inflatable' (Ganfield 2011). The artwork is often seen at fashion shows in a latex head mask with stylised flowing hair, wearing shimmering latex dresses, sometimes accompanied by a white, inflatable polyurethane puppy (Fig. 4.3). I was fortunate enough to interview her (email to author 21 November 2017).

Question: 'Can plastics be a muse for future feminist innovation?'
Answer: 'Plastics, combined with imagination, have the Promethean quality to refashion one's form into anything one so desires.

Fig. 4.3 Pandemonia
2020. (Photo:
Pandemonia)

We have the ability to re-invent the world. The question in my mind is what does one desire? Where do these ideas come from? The pursuit of beauty is not inextricably linked to health but often the opposite. Without realising it we are moving towards the things we are trying to get away from. "Perfection" is a manmade concept: it is alien. Plastics are the embodiment of that concept made real. The physical body has to yield to the rigidity of the idea. Plastics are hard to wear and alien in the environment. Ideals are not found in nature. Prometheus took the flame of knowledge but ended up paying for it.'

Question: 'What does your image represent for the future of feminist innovation?'

Answer: 'Pandemonia's image demonstrates that we communicate through signs. Visually, femininity and masculinity are created out of tropes that can be worn and discarded at will. It shows we can have numerous identities and perform numerous roles. Maybe Pandemonia glimpses at an anarchic freedom where one is free of gender, conventions, society. Through performing, Pandemonia I have been able to travel the world, passing through the invisible walls that restrict us.'

Pandemonia's point about being able to float through restricting invisible walls is in tune with Donna Haraway's famous essay "Cyborg Manifesto". She uses the metaphor of the cyborg to urge feminists to move beyond the limitations of traditional binary gender roles (Haraway 1985). My designs in plastics support such an endeavour. The beret is worn alike by men and women and, as modelled by Agyness Deyn on the cover of a 2006 *Vogue Italia*, presents a gender redefining aesthetic of androgyny. Likewise, James Stopforth's image of a model wearing a yellow visor, taken against a backdrop of waste plastics, presents, in spite of the elegant clothes, a non-binary gender identity (Fig. 4.4).

Beth Buxton, who was instrumental in setting up the shoot, saw her as 'a strange almost AI "worker" in the plastic nightmare' (email to author 18 April 2020). The image was created for and was part of a collaborative exhibition showcasing Wain Ru's architectural concept of a new city borough built entirely from its own recycled plastics and Stopforth's interest in the parallel nature of the plastics world. The concept was to introduce 'a hyper-real part of our world but also a future that may be imminent' (Red Bull Studios 2014).

Plastics Armour

The feminist warrior can often be found in my garments and images. Sometimes they recall Pierre Cardin's outfits of the 1960s, which often suggest the power of plastics as a unifier. Cardin's models look like they are in uniform and their stance is almost alien, somewhat like a cyborg

Fig. 4.4 *Untitled*, 2014. (Photo: James Stopforth styled by Beth Buxton)

vision of a future female doll army. The outfits are also wipe-clean; nothing leaves a mark on this plasticised armour. This reverberates with my ideas about plastics' potential as a tool for the emancipation of women, moulding them into militant, powerful beings, capable of determining and fighting for their own destinies. Cardin's use of plastics is, however, the antithesis of Pandemonia's. Her plastics-wear accentuates her shapeliness, giving her a supernatural silhouette; his outfits transform the wearers into a troupe or army, donning protective hats, gloves, and boots: hygienic, utilitarian and homogenous. Plastics are though at the heart of both aesthetics. Women can choose whichever better suits the identity they wish to project or the protective armour that most accords with their vision of themselves.

Pandemonia's adoption of a plastics persona demonstrates that we can decide who we want to be and how we want to be seen. A form of feminist rebellion is the autonomy to choose your own identity, to reject traditional feminine archetypes created for women and to choose, for yourself, the kind of aesthetic you want to embody. Plastics, as signified by Pandemonia, can enhance the physical manifestation of these rebellious pursuits.

An image that has been influential for me is from the iconic 1982 film *Blade Runner* when Zhora is defying physical boundaries by running through glass to escape Deckard, her strong feminine physique cutting a sharp angular dash through the glass, clad only in a plastics mac.[3] The visual suggestion is that Zhora's plasticised covering moves with her 'replicant' android (synthetic) 'body' through the glass as if the plastics provide a superhuman protective shield. The mac has become her armour. I often reference this aesthetic in my work. A striking example was the finale of my 1995 MA graduate show, with model Michelle Legare wearing a structured transparent mac and inflatable shoulder pads, reminiscent, also, of the black and white 1936 Alex Korda film *Things to Come* based on a story by H.G. Wells. Its materials were plastics sieve filter mesh and PVC with a halo made from a large hosepipe and fairy lights.

Plastics used as armour feature in multiple superhero movies but perhaps most notable and relevant is Tim Burton's 1993 film *Batman Returns*. Michelle Pfeiffer's downtrodden secretary constructs her own PVC cat suit, transitioning from dowdy assistant to an icon of feminine vengeance, Catwoman. This is reflected in how her body moves in the suit; her supple flexibility seems bestowed by the suit's elasticity; but more than this, the character has sewn into this suit the very rebellion she so desires in 'ordinary life' (Fig. 4.5).

In the transformative scene, Selena reaches into the back of her closet and selects a daring PVC black garment that is obviously seldom or never worn, and reimagines the stretchy PVC as a new skin; and in so doing, transforms her personality from anxious and timid to powerful, deadly and seductive. Caroline Evans has explored the association of the female body

Fig. 4.5 Michelle Pfeiffer as Catwoman in *Batman Returns*, 1989. (Photo: marka/eps©agefotostock)

with adaptability, as a kind of malleable plastics maquette that can be re-fashioned as a model for new attitudes and roles (Evans 2013, 215). I find the idea of women using plastics to re-fashion their own ideas of feminist uprising and of future generations of women being influenced by the technological advances plastics have brought into our consciousness especially inspiring.

Plastics can also play a more invasive role in refashioning, seduction, body armour and projecting a 'preferred image' as demonstrated by my friend and peer, Dora Szilagyi. She has what society calls a set of 'fake boobs', 'plastic tits', 'implants', 'enhanced breasts', 'breast augmentation'. This is her personal take on her ownership of her pair of 'fake, new boobs' (email to author 28 November 2017):

Question: 'How do you consider your new boobs? Are they an enhancement of your natural form? Or do you think they give you special power?'

Answer: 'They are definitely enhancements. I have always had big boobs and throughout my life, they disappeared. My body image is pretty unrealistic, so I [confess I have] never had an objective view of my own shape or size. It was a lover who made me aware that my boobs are not as big as I thought. It was not even a question for me to surgically close that gap. They feel more "me" than before.'

Question:	'Do you refer to them as synthetic or plastic?'
Answer:	'Plastic. But mostly I refer to them as "fake".'
Question:	'How might you relate this enhancement to the male gaze? Does this come in to it? Did this enter your head when you were considering this decision? Or do you consider this action a purely personal choice?'
Answer:	'It is interesting because I have not had them done because of one particular person. However by nature most of my motivation comes from a sexual motivation—So, for example, I do sports and work out hard because I am attracted to athletic guys who are into athletic body-type girls. In that way I see the surgery being on the same level/in the same category as working out or eating clean or not smoking, drinking water, using sunscreen—I don't do any of that in particular for the 'male gaze', but all together contributing to the kind of woman that I would like to project to that male gaze. I am quite promiscuous and so the boobs have been, and are, one of my "super powers" in that respect. So far every man I have been with since the operation has been excited about them.'
Question:	'Plastics are controversial and they provoke strong opinions ... do your boobs lead to similar reactions?'
Answer:	'I am very open and non-apologetic about it so, no. It is not often people confront me because my look (tattoos, piercings, strong glasses) is my personality. I also had a breast lift 10 years ago, a mini lipo-suction 7 years ago, implants and another lift 2 years ago. This year I'm planning a bigger lipo-suction for the areas that I can't shift with my workouts. Next things I am thinking eventually another breast lift, laser treatment for cellulitis, veneers and some minimally invasive facial treatments (non surgical). We live in a world where looks are most important, where single women are looked at as a social mishap and where narcissism is an accepted behaviour at any age. I am taking ownership of my body and all

these "improvements" make me feel like an upgraded version of myself. I am not afraid or ashamed to be naked in front of anyone, in any lighting and at any angle because I know that I am making the most of what I have got.'

This is, obviously, an incredibly personal account of a woman and her relationship with her enhancements, and I am sure there are other women who do not have the same opinions of their surgery, their body image, or their supposed superpowers. I think what excites me about Dora is that it is so far removed from the relationship I have with my own body. There is a dichotomy between the male concept of the 'plastic fantastic female' (Kim Kardashian super-inflated curves) versus edgy, minimalist, boyish, intelligent beauty, which I prefer as a designer. But, whatever our preferred body image/shape, we do share similar coats of armour. We all have a coat of armour. We can all choose what our superpower is.

The controversy of plastics in surgery, botox and body augmentation, ties into my fascination with the world's view of this magical substance. It can completely change someone's natural body shape/function: it can enhance an element of your body you are not entirely confident of, or it can replace a missing limb, and that is surely brilliantly exciting? Plastics are a vehicle for both the creation of the mundane and the extraordinary, to quote Barthes again: 'the mind does not cease from considering the original matter as an enigma. This is because the quick-change artistry of plastic is absolute: it can become buckets as well as jewels. Hence a perpetual amazement, the reverie of man at the sight of proliferating forms of matter ...' (Barthes 2000, 97).

Szilagyi's answer in regard to the male gaze recalls the persona of the futuristic über-female Barbarella, the space-traveller and representative of the United Earth government in Roger Vadim's 1968 film of the same name. Portrayed by Jane Fonda, her costumes were designed by futuristic fashion designer Paco Rabanne and included see-through corsets as a form of armour. This relates to the work I did with plastics fabricator and model maker, Kees van der Graaf, aka Mr Perspex, for the US lingerie company, Victoria's Secrets. Van der Graaf (n.d.) is known especially for making torsos and body casts for Alexander McQueen. He also made a body cast of Kylie Minogue, contributing to her dramatic entry from an opening in the floor of the stage wearing a shiny metalised moulded fibre-glass suit of armour as part of her Fever Tour 2002. We were fortunate to be able to re-use it. Our brief was to make a black plastics trench coat that called to mind both Barbarella and Issey Myake's 1980 breastplate as worn by

Fig. 4.6   Fibreglass trench coat designed by Flora McLean for Victoria's Secrets, 2008, Miami fashion show. (Photo: Matt Irwin for *Vogue* Russia)

Grace Jones (Heron-Langton 2020). The trench coat was to appear to be caught in the wind and the surface to appear shiny and wet. The sense of the piece as a protective breast plate was increased as the nipples proved unacceptable to TV in the United States and had to be sanded off (Fig. 4.6).

CONCLUSION

This chapter has demonstrated a variety of ways, some private and some public, that plastics have moulded how people present themselves and thus how they are perceived. My own plastics aesthetic and embodiment of plastics as a muse and constant inspiration is proof that one's destiny, fate, and future can be determined, moulded and sculpted like plastics. Plastics will always be part of my conversation with the world. I am drawn to them as a materials group of artifice and synthetic chemistry. Plastics can be seen as a counterpoint to the idea of the 'natural woman', as a rebellion against oppressive female traits. Women are required to aspire to an impossible 'natural' archetype and embracing plastics is a way of rebelling against it. Plastics are celebrated for their plasticity, for their propensity for infinite change and shape shifting. They can be melted down and also reformed and articulated in many ways, and this appeals to the mercurial nature of many post-modern feminists.

NOTES

1. Story by Daniel Waters and Sam Hamm, directed by Tim Burton, released 1992.
2. Claude Montana's House of Montana was founded in Paris in 1979 and ceased trading in 1997.
3. Story by Philip K. Dick, directed by Ridley Scott. Released 1982.

BIBLIOGRAPHY

Babitz, Eve. 1980. *Fiorrucci: The Book*. New York: Harlin Quist Book: unpaginated.

Barthes, Roland. 2000. *Mythologies*. London: Vintage.

Cummings, Neil. 2004. From Things to Flows. In *The Ecstasy of Things*, ed. Thomas Seeling and Urs Stahel, 322–327. Wintertur: Steidl.

Curtis, Barry. 2014. Dressing for the Future: Speculative Fashion in 1930s Hollywood. In *Fashion Film and Consumption*, vol. 3:1, 47–59. London and New York: Intellect.

Evans, Caroline. 2013. *The Mechanical Smile: Modernism and the First Fashion Shows in France and America, 1900–1929*. New Haven: Yale University Press.

Frieze. 2000. The Future is Orange. Accessed 10 April 2020. https://frieze.com/article/future-orange.

Ganfield, Katia. 2011. Pandemonia Not Just Another Attention-Seeking Tranny. Accessed 10 April 2020. https://www.vice.com/en_uk/article/7bjja9/pandemonia-roy-lichtenstein-inflatable-seven-foot-tall-.

Haraway, Donna. 1985. A Cyborg Manifesto: Science, Technology and Socialist Feminism in the late Twentieth Century. Accessed 10 April 2020. https://warwick.ac.uk/fac/arts/english/currentstudents/undergraduate/modules/fictionnownarrativemediaandtheoryinthe21stcentury/manifestly_haraway_----_a_cyborg_manifesto_science_technology_and_socialist-feminism_in_the_....pdf.

Hawes, Elizabeth. 1938. *Fashion is Spinach*, 289. New York: Random, House.

Heron-Langton, Jessica. 2020. A Brief History of the Breastplate Fashion. *Dazed*. Accessed 10 April 2020. https://www.dazeddigital.com/fashion/article/47465/1/breast-plate-zendaya-tom-ford-mcqueen-ysl-issey-miyake-mugler-grace-jones.

Iley, Chrissy. 2008. Return of the Dandy. *The Guardian*, April 17. Accessed 10 April 2020. https://www.theguardian.com/lifeandstyle/2008/apr/17/fashion.popandrock.

Inflate. n.d. Accessed 10 April 2020. https://inflate.co.uk.

Jaffe, Deborah. 2009. Independence. *Plastiquarian* 41: 2–3.

McLean, Will, and Pete Silver. 2015. *Air Structures: Form and Technique*. London: Laurence King.

Pitchfork. Björg. Vulnicura. n.d. Accessed 10 April 2020. https://pitchfork.com/reviews/albums/20181-vulnicura/.

Poly Styrene—X Ray Spex. n.d. Accessed 10 April 2020. https://www.punk77.co.uk/wip/polystyrene.htm.

Pressman, Laurie. 2017. Panton, Orange Radiant and Fun Loving. Accessed 10 April 2020. https://www.pantone.com/color-intelligence/articles/colors/orange-radiant-and-fun-loving.

Quant, Mary. 2012. *Quant by Quant: The Autobiography of Mary Quant*. London: Victoria and Albert Museum.

Red Bull Studios. 2014. *Plastica*. London: Press Release.

San Francisco Museum of Modern Art. Utilitarian Materials: Scavenging, Stacking and Piling. 2019. Accessed 10 April 2020. https://www.youtube.com/watch?v=kwy7tvUbWgw.

Van der Graaf, Kees. n.d. Accessed 10 April 2020. http://studio-van-der-graaf.blogspot.com/.

CHAPTER 5

Deplastification?

Sebastian Conran

Since first gnashing on a teething ring I have had an over 60-year relationship with plastics, a diamond jubilee of sorts. I have also spent much of my over 40-year design career creating products fabricated from the many synthetic polymers collectively labelled plastics; it could be argued that plastics are the ultimate protean materials group, as the gamut of polymers available today is so wide, all with their own qualities, strengths and weaknesses. They offer an unprecedented choice to the discerning designer and engineer by allowing a great deal of flexibility and scope to incorporate significant functionality into components that is not feasible with other materials, whether for cost or functional considerations. There are also many instances where no other material would be suitable or would be prohibitive due to issues such as weight and corrosion.

As designers we aim to specify polymers in our products under the premise that plastics are not inherently bad materials. It is the irresponsible way in which designers, engineers, manufacturers, retailers and the public use and dispose of them that has led to a rather negative view. Often in the predominant use of single-use products, for example, packaging, water bottles and plastic bags, this criticism is well justified, however, the blame

S. Conran (✉)
Sebastian Conran Associates, London, UK
e-mail: sconran@sebastianconran.com

© The Author(s), under exclusive license to Springer Nature Switzerland AG 2020
S. Lambert (ed.), *Provocative Plastics*,
https://doi.org/10.1007/978-3-030-55882-6_5

89

might equally be shared by manufacturers, retailers, consumers and local waste disposal agencies.

Thermoplastics, developed early in the twentieth century, which melt with heat a bit like hard wax, have been in wide commercial use for about 70 years. Interestingly, manufacturers Addis, once royal warrant holders for their toothbrushes and now a homeware brand more famous for their plastics buckets and swing-bins, started using hygienic thermoplastics as an improved substitute for wood and bone. The German manufacturer of playful and colourful homewares Koisol, which originally started in the Alps where farmers would use the winters to whittle bone as ornaments and souvenirs, also began using the new thermoplastics in the 1940s and 1950s. Their advantages were that they could be injection moulded with great efficiency; they also offered a wide gamut of colours, and were hygienic and safe from the sharp edges found, for example, on tin toys. Although the tooling cost with thermoplastics was quite high, the 'piece cost' was very low; for instance, once you had made the tool, far fewer people needed to be employed to make a 1000 plastics buckets, as opposed to the heavy, dent and rust-prone galvanised, steel ones they replaced.

CHILDHOOD PERCEPTIONS

As a child growing up in the 1950s and 1960s, I had quite a strong relationship with plastics, as many of my toys were made from this colourful ubiquitous material. The varying quality of the different types of polymers was quite apparent to me, with high-quality Britains toy soldiers and Airfix model kits, and at the lower end there were cheap, almost ephemeral toys I could buy with my pocket money from what we called the 'Hong Kong Toyshop' in Inverness Street market in Camden Town.

The first time I recall noticing that not all plastics were equal was when I was about six years old and my father had given me one of the first Lego sets imported. Over time, my Lego bricks would not fit together the way they used to and had become loose, so I glued them with polystyrene glue to make the improbable structures my imagination demanded. Later on, I acquired another set of Lego that behaved slightly differently to my old set and soon learned that although the newer bricks fitted the old ones, they had a firmer grip than the previous bricks. For a start they did not glue together with styrene cement anymore and had a much firmer fit that lasted. What was unknown to me then was that Lego had changed their standard polymer from polystyrene, which suffered from 'polymer-creep' with age, to the more resilient and expensive acrylonitrile butadiene styrene (ABS), which did not distort with time as much. I still never cease to wonder how Lego have made so many billions of bricks so accurately that snap together so precisely whenever and wherever they were manufactured. This is a good example of a benign business that could not exist without plastics and of exceptionally high-manufacturing standards.

So even to an inquisitive six-year-old child who loved taking things apart to see how they worked, there was quite an apparent difference between polymers such as polyethylene (PE), polypropylene (PP) and polystyrene (PS). One key thing was that PS could easily be painted and glued with PS cement, whereas the slippery finish of PE and PP could not; the other was that PS was capable of much better detailing and finish, however, PP could bend without breaking. When I later discovered polycarbonate, it seemed like a magic material because of its robustness and 'ability to stop bullets'.

Later, at primary school, I became obsessed with building Airfix model aeroplane kits made from polystyrene and then meticulously painting them with Humbrol enamel paint. These were creations of pride and there was significant competition amongst fellow modellers in school. I learned a great deal about the different types of plastics, making things and about aeroplane design and it was not long before I started making my own designs and inventions too. My childhood ambition was to be an inventor and without doubt my encounters with plastics Lego and Airfix as well as Meccano, made of metal and held together with nuts and bolts (more grown up and versatile), enhanced the quality of my childhood and encouraged my imagination and aptitude for making things.

At secondary school I excelled at maths, physics and chemistry, all subjects which I use daily in my design work, however, I spent most of my free time in the metalwork or woodwork shop. Plastics raw materials were not as available in schools in the late 1960s as they are today. It was this desire to invent and make things that drove my decision to study industrial design engineering at what is now Central St. Martins and led to a career in product design. Fortuitously, my first girlfriend's father also ran an innovative injection-moulding factory in Slough, which inspired my love of polymers too.

Professional Practice

Although there are many varied factors to consider in the design process such as people skills, engineering-integrity, cost and aesthetics, it is the purchasing customers and end users who are the ultimate judges of success, not just peer press acclaim. Product design may well be a commercial activity, but it is also one that needs to engender lasting appeal that satisfies not just the client, manufacturers, retailers and customers, but also contributes to the end user's life for many years. The significance of this is discussed in greater detail in Chap. 12.

My dictum of 'form follows fabrication' is important as it is the crux of why people use polymers today, it is the ability to produce quite complex components with a great deal of function and a high degree of precision and finish in a very short time, whilst not requiring a great deal of craftsmanship and often delegating the factory workers to a lesser skilled assembly activity. What I did not know at the time was how many hundreds of products I would design during my career fabricated from a plethora of materials including polymers. During this time, I discovered that the choice of polymers for a particular component used in a design was critical, not only to the way products looked and felt to the touch, but perhaps more importantly to how well they functioned, and behaved under stress, their aptitude for failure, as well as their impact on manufacturing simplicity and cost.

The advent of the late twentieth century as the plastics age, not unlike the Bronze Age, can almost be compared with the invention of the printing press, which allowed Johann Gutenberg to produce hundreds of bibles in significantly less time than it would take a monastery full of monkish scribes to produce using quills on parchment. A key similarity is the significant upfront investment required to buy a printing press or an injection-moulding machine, and to create the printing plates or mould tool. As with books, design is the legacy of the thought of the writer/designer, and equally the legacy of design is the thoughts of the users; the first being the creator's rigorous thinking process expressed as a tangible object, followed later by someone's mental perception of the experience of using said design.

When designing products using polymer components, it is essential to get a thorough and rigorous understanding of the specific polymers to be used, their capabilities, qualities and production process and the toolmaking and design; to imagine how the hot polymer will behave when it flows around the tool in production and also how the final moulding will be extracted. Often, discerning design choices, made throughout the creative design development process, can have a significant impact on the ease of manufacture, reliability and cost of the item. On the face of it, the injection-moulding process is the designers' dream come true. Once the product has been designed and tooled up, the production costs are minimal compared to traditional fabrication techniques, and materials/component sourcing is much less complicated. However, all too often plastics are cost-engineered to give a product just enough lifespan to make it serviceable for a short period of time. In contrast, I like to design with traditional natural-feeling materials such as glass, wood, ceramics, and metal only using polymers when they are the best, rather than cheapest, design solution to fabrication.

As product design consultants, we feel our core objective and duty is to create 'value' within our client's business. In a commercial context, 'value' is the engine of trade, without it there is no exchange of goods and money. This is a rather rational definition of 'value' as the perception of fair exchange, however, we focus on the 'customer' experience. There is another more nuanced and emotional meaning of 'value' when it relates to how much we cherish or love something such as a favourite pair of shoes or our pets. This definition outweighs the actual cost and has more to do with the 'user experience'. Whichever way you look at it, you cannot get a user experience without a positive customer one first: someone has to buy the product in order for it to get used and experienced. For this reason, I always think about how a product is going to be sold and perceived by the purchasing customer as well as by the person who will be owning or using it.

Crayonne—In the early 1970s, I was working as an apprentice in the Conran Associates design studio, when they were designing a range of innovative homewares to be manufactured by Crayonne Ltd, a division of Airfix. These beautifully designed products were destined for a premium market and were injection moulded from ABS. Some had thick sections of over 5 mm and narrow draw angles, which was possible as ABS shrinks less in moulding than, say, the cheaper polypropylene with which it would not have been feasible to make them. They launched well, but after the oil crisis the price of polymers increased dramatically and together with the long cycle time required by the large wall thicknesses to cool, these elegant products were no longer economic to make.

Mothercare—In 1981, I was recruited as head of product design for Mothercare. At that time, it had little visual cohesion and felt rather like a chaotic camping shop for mothers. It provided me with what would have been almost a dream challenge for any product designer: hundreds of items that needed to be redesigned with a common visual style that was feminine, hygienic, useful and desirable. Internally, we called the visual style employed 'Soft Tech'.

Much of the non-furniture merchandise such as baby baths and potties were made in the United Kingdom from hygienic, easy-to-clean and inexpensive plastics such as polypropylene which was durable and kept its finish well. Exceptions included items like feeding bottles, which were injection-blown polycarbonate with silicone or latex teats, and printed PVC sheeting, as it was water and soil-proof, which was used on changing mats and for linings of cots and pushchairs.

Pushchairs are a significant product not only because of their cost but also they are probably the only baby product predominantly bought by males, most likely because they have wheels. I recall my first visit to a key pram and pushchair supplier/manufacturer called Restmor in Letchworth, and being shown around their metal-bashing facilities and their brand new steel-chroming plant. After this we had a meeting when I discussed my vision for fabricating their future wheeled products from engineering polymers. There was immediate alarm at this idea, 'plastics are ok for toys!' they exclaimed, 'buggies will break if we use plastic parts and we have enough problems with damaged-goods returns already'. To be fair, their factory was set up and completely optimised for chromed steel tubing, so this was going to involve substantial investment.

Eventually, with the aid of a Rossignol ski boot to demonstrate how tough polymers could be in adverse conditions, we were able to convince them that this would be possible in collaboration with key London-based plastics supplier, Bissel. The result was the development of the highly innovative and British Plastics Federation award-winning VIA pushchair. It was lighter, more contemporary looking, had many more useful features and benefits and was versatile. It could fold with one hand (the baby being held in the other one) and was significantly less prone to damage in use than their previous pushchairs. Most importantly, it sold in huge quantities, over six times the product it replaced, and delivered significant bottom-line profits. Key to this success was that we had started with a clean sheet of paper and employed engineering polymers, mainly nylon 66 for structural components with some over-moulding on the aluminium front tubes, requiring the metal components to be placed inside the tool and the polymer moulded around it. This delivered greater value to the consumer, not by making it cheaper, as in fact it was 20% more expensive to purchase, but by offering a much better and more desirable product that was more focused on the whole experience of owning and using it. This was only possible because we had invested in the design and tooling of polymer components that increased functional behaviour and performance. We went on to successfully design many other buggies in the same vein and as a result fundamentally influenced the future pushchair industry (Fig. 5.1).

Apart from the obvious designing of Mothercare merchandise, one of my tasks was also to look at garment hangers, which turned out to be quite intriguing. The challenge was that garments from different manufacturers came in on visually uncoordinated hangers, so there was little choice but to start from scratch. Their redesign used an innovative-for-its-time 'C' profile, which made the polystyrene hangers look more substantial and visually appealing than the more utilitarian look of the H profile hangers that had gone before, whilst using less plastics and achieving a similar high cycle-rate. This was only possible with inexpensive polystyrenes, as the alternative polypropylene shrinks significantly whilst cooling, creating unsightly distortions in the finished product such as shrink 'dimples' and irregular bending. Polypropylene also suffers from 'polymer creep', which over time means that it is more 'plastic' and will memorise a distortion, unlike the stiffer polystyrene, which is more resilient and will bounce back to its original shape. The thinner the material the quicker it cools, and thereby the more productive the injection-moulding machine can be within a given time. This is why designers in plastics are always under pressure to use as little material as possible to do the job well.

M&S—Soon after I set up my own independent studio, one of my earliest clients was a Swedish company called Karner who had noticed the innovative work I had done with the Mothercare hangers. My first project for them was to design a knicker hanger for Marks & Spencer who was one of their customers. The result was nicknamed the 'Centipede', as it had little legs protruding from underneath the cross bar so that different hip sizes could be uniformly displayed. It needed to be an H section to counteract the constant force of the elastic waistband and we used the stiffer polystyrene for rigidity and recyclability, and pared the design down to use the minimum of material. This was the first time M&S had hung their undergarments and like-on-like sales went up 14%. Thankfully, as

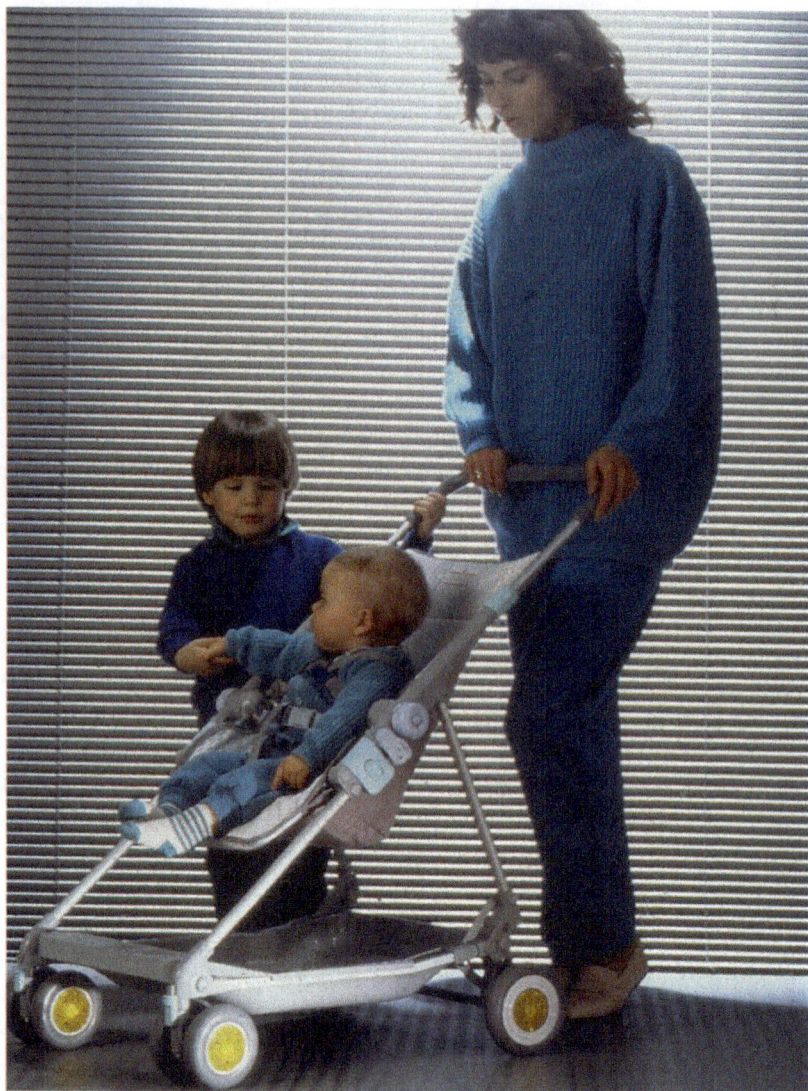

Fig. 5.1 *Via Pushchair*, lead designer Sebastian Conran for Mothercare, 1984. Photo: Conran Associates/Restmor

they sold around 30 million pairs of knickers a year (over the last 34 years this equates to over a billion), M&S has an efficient recycling system so they have not all become landfill. Following this success, I found myself designing many more innovative jacket, dress and skirt hangers as well as other technical packaging such as jersey frames. I was intrigued to find that the garments are put on the hangers in the factory, and that throughout the shipping process the metal hanger hooks are the only things that come into contact with the shipping system—a bit like railway carriages along a train track.

Spud U like—In the late 1980s, my studio was commissioned by Spud U Like to design some disposable cutlery; there was nothing on the market that would decently perform the particular function of eating baked potatoes. Our design solution consisted of a knife and 'spork', a hybrid of a spoon and fork. Moulded in dark green polystyrene, the construction harked back to the successful 'C profile' we had used on the Mothercare garment hangers. They were designed to be compact and sturdy to use, as well as employing as little material as possible and giving them a spark finish so they felt less plasticky in the hand. The client and his customers were delighted with the result and it was shown in the Design Museum as an example of good design of its type. Although theoretically these 'Sporks' were fully recyclable, they had to be made from virgin food grade polymer rather than the recycled material we used in clothes hangers.

In response to the completely valid concerns that after 70 years of use we are drowning in un-compostable disposable thermoplastics, especially from packaging, designing single-use disposable products is naturally something I have always had qualms about, principally due to their environmental impact. Thus, our aim is always to fulfil the brief using as little material as is feasible within out remit thereby restricting consequential environmental damage—as a designer you have limited influence on the commercial requirements of your clients.

Logically, and discussed at great length in Chap. 8, it does make more sense to use oil to make useful products that can then be 'terminally recycled' to produce electricity, rather than burn oil in power stations. This approach is favoured in Denmark (home of Lego), where they have the recycling infrastructure to support it and interestingly a terminal recycling plant located next to their iconic Copenhagen opera house.

More recently, I have been working on compostable water bottles, which have their own challenges as they cannot be recycled with normal polymers as they would adulterate its integrity and physical characteristics—imagine a railway sleeper made from recycled material and with age some of the polymer degrades.

Anywayup Cup—In the mid-1990s, a colleague showed me a newspaper article about an innovative toddler cup and its patented spout valve that prevented it from spilling when dropped. She thought we could make a better job of designing its rather generic appearance as I had considerable expertise from my experience at Mothercare. So, rather cheekily, we contacted the inventor, Mandy Haberman, who agreed that it could be improved but not quite for the same as our reasons. It transpired that she had commissioned another reputable design company to undertake the original development, and although the result looked passable, it had functional issues with leaking, not from the spout, but from between the lid and the base. We found that the key issue was that the hard (and expensive) polycarbonate base would not seal properly with the hard ABS lid using the bayonet closure employed. Our conclusion was that it would be better to use different materials, polyethylene (PE) and polypropylene (PP) as their molecular structure is different: it is often difficult to get a good seal between similar materials as they tend to 'bind' together (a bit like Plasticine) rather than slide over each other.

Our proposal, thus, was to use a Tupperware-like snap-on lid made from polyethylene with the patented co-moulded silicone valve in the spout and a translucent tactile spark-finished polypropylene body. As the material cost of PE and PP is about a quarter of that of polycarbonate and ABS, there was an overall saving on manufacturing cost of about 25%. We also restyled the whole cup in a way that was innovative looking and more 'of its time' than the previous model resulting in a 4000% sales increase. The design was so distinctive that the client trademarked its appearance which effectively protected it against attempts of plagiarism (Fig. 5.2).

Smart Café—Later in the 1990s, we had the idea for a hybrid between a cafetière and a coffee mug and designed the award-winning *Smart Café* (I wanted to call it the iCup). It has now been in continuous production for over 25 years; it still gets 4/5 stars on Amazon and sells pretty well. Fabricated from mainly ABS mouldings ultrasonically welded together and an over-moulded metal filter sieve, it would be nearly impossible to make with traditional materials and achieve the same functionality, let alone cost. We have explored fabricating the concept in different ways and from alternative materials such as metal, ceramic or glass, but never been able to achieve the same usability due to the very high tolerances required that only injection moulding can achieve (Fig. 5.3).

Fig. 5.2 *Anywayup Cup*, lead designer Sebastian Conran for Mandy Haberman, 1997. Photo: Sebastian Conran Associates

When it was launched, I suggested to my father, Terence Conran, the founder of Habitat and the Conran Shops, that he might like to stock it in the Conran Shop: 'I don't like drinking coffee out of plastic', was his slightly haughty response. This sort of prejudice is not unusual when people are being offered a 'plastic' coffee mug (personally, I prefer ceramic mugs too). They often complain that it tastes 'plasticky', but we found from blind tastings that the ABS really did not affect the flavour. It may be just human perception. This said it is important to wash any plastics that come into contact with food thoroughly to remove any residual mould release lubricants before use.

Plastics beer glasses in pubs are subject to the same sensory prejudice, even though hundreds of people arrive in A&E every year with injuries from broken glass ones. Logically, clear polymer ones made from polycarbonate, SAN or equivalent would work better. They have been trialled in tough transparent polycarbonate and are available, but not in general use except in special circumstances. A study was undertaken when I was involved with a collaboration between the Home Office and the

Fig. 5.3 *Smart Café*,
lead designer Sebastian
Conran for Zyliss, 1998.
Photo: Sebastian Conran
Associates

Design Council in 2010 as part of the 'Design Out Crime' initiative (Design Council n.d.). Although all the logic and tests demonstrated this should be done, sceptical drinkers complained that the beer does not taste the same. I suspect it might also be the 'flex' and lack of heft that puts people off.

So what is it that makes people react so seemingly irrationally? In short, it is the way that contemporary plastics feel in relation to the sense of touch, or rather relating to the perception and manipulation of objects using all the senses. We call this the 'haptic quality' of products and finishes; it derives from the Greek 'haptikos', to touch or grasp. An unexpected example of this occurred when we were working on the interior design of the last British Airways Concorde: weight was an important issue, so we developed and tested some very beautiful lightweight titanium cutlery only to find that passengers did not like them, as their lightness made them feel like plastics.

Equilibrium—Another example of haptics at work relates to a set of traditional kitchen 'balance' scales we were asked to design and re-imagine in a modern manner in the mid-1990s. We did a fair bit of discussion, brainstorming and sketching in the studio but I had a lingering feeling that we did not have quite the right answer yet. Then, on a family weekend, I found myself on the pebbled Chesil Beach in Dorset and started thinking that we could make the weights in pebble shapes. A while back, it would have been difficult to design an irregular shape and to precisely determine its weight, let alone draw it up so a manufacturer could make it; however, we had just started using Computer Aided Design (CAD), discussed in greater detail in Chap. 3, which enabled us not just to create amorphous shapes, but also to know exactly how much each would weigh in a given material. The form of the pebbles was the 'Big Idea' that influenced the form of the scales themselves (Fig. 5.4).

Originally, we were going to make the body of the scales out of cheaper polymers, but they felt a bit light and insubstantial so we chose to make the upper carapace of chromed aluminium, and the base and weighing trays from ABS and polycarbonate. They were a big success. John Lewis used huge images of them in the publicity for their new kitchen department but the department buyer was perplexed as to why they sold so well as digital scales were easier to use and cost much less. One day I stood in John Lewis and from a distance watched the display of the scales and saw customers pick up the weights and roll them in their hands—the body language indicated pleasure, and they would pick up the box and go to the till.

In 1988, Apple launched the iconic 'Blueberry' iMacs with their colourful transparent organic forms displaying the inner works of the cathode ray tube, something that was only possible using engineering polymers such as polycarbonate and importantly played to the innate qualities of the material. Their striking appearance had a huge impact not just on the design of computer equipment and accessories but also on many other aspects of product design in polymers, for instance, our Anywayup cup. It soon seemed that everywhere you looked there were colourful transparent enclosures around everything from data drives to vacuum flasks.

What followed was even more intriguing, as the next version of the iMac had a LCD flat-screen on a polished stainless steel articulated arm

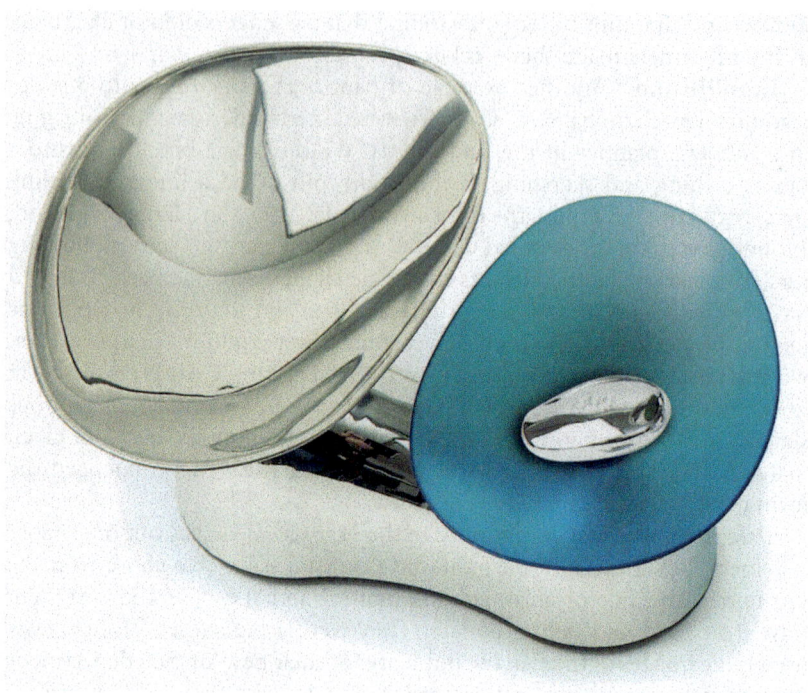

Fig. 5.4 Scales, lead designer Sebastian Conran for Equilibrium, 1998. Photo: Sebastian Conran Associates

coming out of a translucent matt-white polymer hemisphere that housed the computer, rather like a sleek space-age Anglepoise lamp. It gave a real perception of quality. This use of polished steel with white plastics was further employed on the first iPod where its polished stainless steel back gave what would have otherwise been a simplistic 'brick of plastic' a sophisticated aura that eventually evolved into the iPhone, which is undisputedly one of the most successful product lines of the early twenty-first century.

When I look around my own kitchen, I find that my favourite utensils are in some cases 40 years old and generally made of traditional materials such as ceramic, glass, wood or steel. Although over the years I have designed and purchased many plastics kitchen gadgets myself, few have survived and those that I still have are kept in a drawer rather than out on display; maybe this is because plastics just do not age well. The key exception are dishwasher-safe knife handles, where using polymers is inescapable, so we 'linish' off the plasticky finish with fine wire mops to give a matt 'brushed' surface after the over-moulding process; this is achieved by inserting the knife blades into the injection-moulding tool and forming the hot polymer around the metal component giving a strong and hygienic assembly.

Ceramics and glass do not scratch and keep their appearance throughout their lifetime until they get broken or attacked by a maladjusted dishwasher. Wood and stainless steel on the other hand can age gracefully. The slightly singed patina of a well-used wooden cooking spoon is not a reason to get rid of it. With thermoplastics, however, burns and scratches can appear quickly and are usually less than attractive; colours can be fugitive and fade in daylight, surface finishes start to 'bloom and flake'. Plastics are more subject to degradation in use due to several factors: stress weakening, polymer creep, plasticiser migration and oxidisation. When this happens, depending upon the polymer, it results either in failure or a marred finish. For example, I have been wearing acrylic-formed spectacles for almost 60 years and find that over a few years they get scratched and bloom appears on the frame that is difficult to remove, whilst the glass lenses, unless broken, are as good as new.

As human beings, we perceive the world and objects through our senses: living, eating and cooking are sensual activities. Designing the products we come across and use every day is about much more than just joining the dots between usefulness, usability, functionality, appearance, finish, fabrication, materials, cost, brand, and so forth. It is also about providing a personality and behaviour for things and an experience for the user that will be pleasurable and satisfying. It means imbuing designs with a sort of inanimate soul, which expresses the thought that has gone into both designing and making the objects. This is a notion I have found the Japanese inherently understand as it is part of their culture, they were also one of the first nations to embrace the use of plastics in their new post-war factories built from scratch without legacy funding.

Deplastification is an invented term that I use when I want to design and make things feel and appear less 'plasticky'. This does not necessarily mean not using polymers, but using them in a way that adds perceived quality to the resulting product design as opposed to detracting from it; therefore, it is important to users that their products do not feel too cheap, 'plasticky' and unpleasant to touch and smell. Often this negative and flimsy tactile nature of polymers can be masked by in-tool spark finishes and thick wall sections as well as over-moulding or post-moulding techniques such as hand-finishing, painting, covering with cloth, Physical Vapor Deposition (PVD) coating to give a thin metal finish or other finishes to improve the haptic and sensual tactility for the user.

Nissan—A good example of the deplastification approach is provided by when we were commissioned to reimagine Nissan's urban city car the *Cube*, as a display 'Show Car' for the Tokyo motor show in October 2003. In February 2003, we had been briefed that our design solution should use their existing body-shell and core mechanicals, but that we could do what we chose with the interior and trim to give it a premium 'Conran' brand experience. The Conran brand values include: aspirational, original, accessible, elegant, eclectic and useful. These may seem appropriate for the Conran Shop but how does it relate to a mass-market manufacturer's compact car?

Our initial impression of the *Cube* was that it was already very functional, thoughtfully designed and useful; however, there was a lingering impression of a 'plasticky' feel and smell to the slightly generic interior, which also had a proliferation of ersatz leather embossing to the mid-grey plastics, which increased the sense of unexciting cheapness.

Our solution needed to have some 'wow factor' but, nevertheless, be commercially feasible. We achieved this partly by using leather seats and anodised aluminium highlights, but also by using an authentic finish that was true to the inherent characteristics of the polymer parts such as polyurethane that was darker and texturally more akin to a suede without trying to mimic a leather. Although both interiors were largely made from synthetic materials, it was the attention to the haptic feel and sensual experience that made our version more aspirational (Fig. 5.5).

At the start of the project, the intention was that it should be a promotional activity and not go into production; however, a limited edition of 5000 were manufactured: they sold out immediately and the residual value was much higher than that of the original Cube with which we started.

Universal Expert—More recently, we have collaborated with retailers and manufacturers across the globe and launched a 180-piece collection of functionally innovative kitchen goods called Universal Expert. The collection's key marketing messages were: 'It's not made of plastic' and 'why has nobody done this before'. Predominantly, in its design, we used

Fig. 5.5 Interior of the *Cube*, lead, designer Sebastian Conran for Nissan, 2003. Photo: Conran & Partners/Nissan

'traditional materials' such as ceramics, glass, stainless steel and wood. However, we found that we had to use some polymers such as flexible silicone rubber in certain places to improve functionality, hygiene and usability, as we could not find any other available materials that could do the task as well.

REFLECTIONS

In our 'global village', a design must not be merely good but outstanding in order to succeed. When I see or use an outstanding product I may think 'I wish I had thought of that' and I will respect the person who has created it, whether they are a design professional or not. There are some designers, notably Dieter Rams working for Braun and Jony Ive at Apple, that apply an almost formulaic visual signature to their work, with, interestingly, plastics playing a significant part in both cases, whilst always seeming fresh, and other great designers such as Achille Castiglioni, whose designs often foreground the material of which they are made, whether marble, steel or plastics, or Charles Eames, who designed the first plastics chair from fibreglass, whose work is recognisable more through its lateral thinking mindset than appearance. To my thinking, both approaches are valid so long as they are satisfying to use and put 'a smile in the mind' and they design them responsibly.

To me, outstanding design is the legacy of intense imaginative thought and vision, combined with skill and craftsmanship; it is also the ability to anticipate tomorrow's challenges, today, particularly through the choice of materials. In turn, the legacy of design must be a satisfying user experience in which the product is enjoyed and cherished and not sub-standard and destined for landfill, which is often the case in a world where, cynically, products are rushed to market with corners-cut including safety features, to satisfy the company's share price.

My belief is that one must thoroughly understand how a product will be made and experienced in its context before even putting pencil to paper. Rigorous process and close collaboration are also important to ensure a successful outcome, and we feel that great ideas are often created and nurtured through discussing projects with clients, manufacturers, marketers and end users, as well as undertaking thorough research into the market, materials and production methods, all of which can influence the narrative of the design vision. Designing a complex product can feel like keeping a 1000-piece jigsaw in your brain, each piece a different issue, using a process of continuing to push to fit them together in new and different ways to eventually get the result required. And every day you discover a new problem or a new opportunity, to fit these things together a little differently.

Great design is simple but not predictable; it is only obvious after the event. So the amount of love and care you put into a design project is always apparent. Even if people are not conscious of it, they can sense when you have paid attention to every little detail. Perhaps my personal attitude to material selection has changed a bit over the years, however, my general ethos has always been: 'Only use polymers when it is the best material for the job, not the cheapest alternative'.

Colour, form, tactile qualities and functional behaviour are just some of the ways we emotionally engage and experience objects and environments. And as bees are attracted to flowers, it is the spiritual nourishment that design offers that can be so important. The part played by plastics in this spiritual nourishment is paradoxical. They are frequently integral to the design process yet deplastification is an important consideration in achieving a product that will be cherished.

Bibliography

Conran, Sebastian. 1986. *My First ABC* (Illustration). London: Atheneum Books.
———. 1987. *My First 123* (Illustration). London: Atheneum Books.
———. 1989. The Big Wheel. In *From Matt Black to Memphis and Back Again*, ed. Deyan Sudjic, 237. London: Architecture Design and Technology Press.
———. 1990. *The Amazing Umbrella Shop* (Illustration) London: Sidgwick & Jackson.
———. 1997a. *Contemporary Furniture*. London: Soma Books.
———. 1997b. *Castle Howard*. York: Castle Howard Estates.
———. 1998. *Contemporary Lighting*. London: Soma Books.
———. 2005. V = [D+Q+B]/P [Creating Value]. In *The Art of Plastics Design*, 2–4. Shrewsbury: Rapra Technology Ltd.
Conran, Terence. 2016. *My Life in Design*. London: Conran Octopus.
Design Council. n.d. Design Out Crime. Using Design to Reduce Injuries from Alcohol Related Violence in Pubs and Clubs. https://www.designcouncil.org.uk/sites/default/files/asset/document/design-out-crime-alcohol.pdf.
Fiell, Charlotte and Peter. 2005. *Designing the 21st Century*. Koln: Taschen.
Hamlyn, Helen. 1986. *New Design for Old: An Exhibition of New Products Designed to Help Older People Stay Independent at Home*. London: Helen Hamlyn Foundation and Conran Foundation.
Krivine, Andrew. 2020. *Too Fast to Live too Young to Die: Punk and Post-Punk graphics 1976–86*. London: Pavilion Books.

Lefteri, Chris. 2006. *Plastics 2: Materials for Inspirational Design*. Switzerland: A RotoVision Book.

Newson, Alex, Eleanor Suggett, and Deyan Sudjic. 2017. *Designer Maker User*. London and New York: Phaidon.

Sebastian Conran Associates. 2020a. Celebrating a Decade of Design. http://sebastianconran.com/10th-anniversary. Accessed 25 May 2020.

———. 2020b. iCon for Concorde. http://sebastianconran.com/icon-concorde. Accessed 25 May 2020.

Walker, Stuart. 2006. *Sustainable by Design: Explorations in Theory and Practice*. Sterling, VA: Earthscan.

CHAPTER 6

The Imperfect Aesthetic

Stefan Lie, Berto Pandolfo, and Roderick Walden

Manufactured objects are an integral part of society. We interact with objects every day and in doing so there are numerous decisions made during the object-user exchange. Desmet and Hekkert (2007, 59) suggested that there are three levels of human-product interaction (HPI); aesthetic, meaning and emotional. At an aesthetic level, it is how an object can engage with one or more of the user's senses; the experience of meaning level is a cognitive process where an assessment of the significance of an object is made, and the emotional experience level relates to the feelings and emotions that are a consequence of the interaction. Norman (2004, 63–89) suggested that there are three levels of basic human processing: visceral, behavioural and reflective; an over-simplified but nonetheless useful guide to understanding human behaviour and particularly relevant to this context of the object-user exchange. The visceral level of processing is the most basic, assisting in making decisions about, for example, the environment being good or bad, safe or dangerous. The behavioural level is largely subconscious and is about learned skills like playing football or laying bricks, and the reflective level is where conscious thought occurs, where deep understanding is developed and reasoned and conscious

S. Lie (✉) • B. Pandolfo • R. Walden
University of Technology Sydney, Sydney, NSW, Australia
e-mail: Stefan.lie@uts.edu.au

© The Author(s), under exclusive license to Springer Nature
Switzerland AG 2020
S. Lambert (ed.), *Provocative Plastics*,
https://doi.org/10.1007/978-3-030-55882-6_6

decision-making takes place. Regardless of the level of object-user engagement and whether it is conscious or subconscious, whenever we interact with an object we do so in a particular way.

Narrowing the focus of the object-user interaction to particular criteria, which are then observed over a predetermined period of time, may provide a way to identify how, in this case, the criteria of the aesthetic value in relation to the object-user interaction with plastics objects, has significantly changed. Since the emergence of man-made plastics materials in the mid-nineteenth century and the recent proliferation of plastics used in 3D printing, plastics have been a constant participant in the innovative and technological discoveries that have occurred throughout this period. However, the relationship and understanding by manufacturers and fabricators of plastics as a material has significantly changed. Analysing this period in the context of the *material—object—user* experience, may provide a better understanding of plastics beyond their physical attributes. The observations identified between the mid-nineteenth century and today can be categorised into four stages; plastics as a substitute material, plastics as a glamorous (glamour) material, plastics as a high-performance material and plastics as a craft-based material.

Four Stages of Plastics Evolution

The first stage of plastics evolution which broadly relates to the mid-nineteenth century is defined by its discovery and use as a substitute material. Plastics have a much shorter history with respect to the more traditional materials such as timber, glass and leather. Today, however, they occupy their own clearly identified and understood position in the landscape of available materials. The first plastics were discovered and utilised as an alternative to what was pre-existing, used as 'merely a substitute material' (Catterall 1990, 67), and early plastics were employed to replace other scarce or costly natural materials (Sparke 1999, 202). Parkesine, the first man-made plastic from 1862, was used in jewellery items such as bracelets, earrings and necklaces and made to look like ivory, coral or amber (Kaufman 1963, 26). During the latter part of the nineteenth century, material scientists were encouraged to find a replacement for ivory which was used in billiard balls, a pastime that had become increasingly popular. The exotic material ivory was sourced from the tusks of elephants and was still readily available. Producers of imitation ivory billiard balls material made from cellulose nitrate, advertised them as an opportunity for consumers to purchase a product otherwise only available to the wealthy.

The scarcity of tortoiseshell which was prized for its use in decorative hair combs and pins made it expensive. Even the horns from cattle that were used as a tortoiseshell substitute became difficult to source, and again cellulose nitrate was used as an alternative (Meikle 1995, 17). In 1928, a thermoset material Catalin emerged as a colour-fast and hard-wearing material that achieved immediate success, being described as combining the glitter of precious stones with the toughness of metal. Methyl methacrylate commonly known as acrylic was prized for its crystal-like clarity and widely adopted to replace other more expensive and fragile materials such as crystal and glass.

Decorative laminates were developed in the twentieth century and used as a substitute for timber by manufacturers of kitchen, dining furniture and building interiors. The laminate companies were capable of producing surface finishes that resembled natural materials such as timber and stone. This ability to replicate another material was also used in the automotive sector when timber dashboards and steering wheels, once crafted by skilled artisans, were eventually replaced by artificial plastics alternatives. Engineered stone is a composite material developed in the late 1960s and used in place of real stone or marble in walls, flooring and countertops. Composite plastics such as fibre glass was perfected by the 1950s and used as alternative materials in aircraft components, car bodies, boat hulls, roofing, sporting equipment and furniture (Mossman 2008, 139).

Many of these examples forced the perception that plastics were materials used in place of the original and therefore an inferior product. Making plastics look like timber or marble perpetuated a negative stereotype. It was not until mid-twentieth century designers such as Henry Dreyfuss designed objects made from plastics that displayed integrity and uniqueness that a shift began in the meaning of plastics away from impersonator towards the establishment of their own set of unique values.

Stage two of plastics evolution is characterised by its ability to produce high gloss surface finishes elevating them into a position of glamour and striking visual appeal. Mid-twentieth-century design and manufacture were challenged by two opposing ideals. The first was an inherited pragmatic and conservative approach that encouraged honesty in material use and clarity of design message, an ideology epitomised by the modernist mantra 'form follows function'. This approach became a barrier to many in the creative industries and eventually inspired a rebellion against these rigid conventions, allowing an opposition to occupy a second more individual and expressive position. This subject is discussed from the consumer perspective at greater length in Chap. 12.

The partnership between design and industry played a significant role in highlighting the functional potential of plastics in society and it also countered the growing perception that plastics were inferior to natural materials (Sparke 1990, 8). Regardless of the ideological standpoint, what began to emerge was that plastics had identified their own 'sincere image' (Manzini 1990, 134) rather than the representation of something else. This transition, however, was not straightforward. Consumerism and large-scale manufacturing required plastics to be fully integrated and for this to occur, risk became a factor. For example, to manufacture the Sottsass designed Valentine typewriter or the Zanuso and Sapper designed Algol 11 television (Bosoni 2008, 96, 113), significant investment in tooling and commitment to high volume production represented a serious financial and commercial risk. The perseverance and optimism of designers and industry leaders prevailed to ensure plastics would become accepted as materials with their own unique set of economic and cultural qualities (Manzini 1990, 134).

From when they were first used as a visual replacement for ivory and other precious materials, plastics were always associated with glamour and luxury. The introduction of nylon stockings, which replaced silk that was delicate and wool that was scratchy (Fenichell 1996, 136), instigated consumer hysteria that escalated into the 'nylon riots' (Spivack 2012). Nylon stockings were luxurious and glamorous because they were functionally superior to previous products and they were flattering for women. As this was occurring during World War II, they were hard to find which meant they were a highly sought-after product. Plastics made front-page headlines and had become news for the masses.

In the 1960s, the plastics characteristics of colour intensity, flexibility, transparency, flowing form and high gloss surfaces became industrially achievable using a variety of processing methods. It was not long before these traits were being exploited in furniture by designers such as De Pas, D'Urbino and Lomazzi, (Annicchiarico 2001, 74) and garments by Rabanne (Kamitsis and Rabanne 1996, 122–132). This represented an optimism that was symbolic of the time and continued into the 1970s when designers such as Sottsass and Bellini extended the use of colour, style and form in their designs of utilitarian objects for the home and office (Bosoni 2008, 113, 129). Plastics were now prized for their unique ability to be visually impressive. The next challenge was for the plastics material group to confront other material groups. Could plastics be competitive in terms of physical performance?

Stage three of plastics evolution became evident towards the end of the twentieth century when they began to be utilised in precise and high-performance contexts. Plastics had evolved from materials prized for what they represented to materials desired for their performance. Around the turn of the twentieth century, plastics became ubiquitous, occupying functions across the spectrum of society, from the trivial bottle top to the serious human prosthetic. This shift is reflected in the broader cultural acceptance of plastics. Contemporary society is less likely to judge plastics because of what they are or what they are attempting to replace, rather they are accepted and judged on their merits, like most other materials. With plastics reaching this level of acceptance and their continually expanded use, it is interesting to ask, where are the extreme applications and what are the limits of possibility? Increasingly, plastics occupy roles and functions which other materials cannot, or more significantly, would be considered inferior to plastics.

Measuring the performance of a material is dependent on factors such as context and parameters of use and one metric that can be applied to assess plastics is physical size. The first plastics were used for small objects such as brooches, hair combs and cutlery handles. With the development of different plastics and processing methods, the size of objects increased to include items such as chairs, car dashboards and kayaks. This evolution continues, rotational moulding is a processing method that allows for very large parts to be manufactured such as single piece boat hulls and water tanks the size of small buildings.

Apart from the very early natural plastics, the vast majority of plastics were discovered either accidentally or intentionally in scientific laboratories. This environment of control and rigour has enabled continuous investigation to occur in laboratories across the world resulting in thousands of plastics types and varieties. Some plastics have been developed with a very narrow band of characteristics and their use reflects this. For example, Polyphenylene Sulfide (PPS) can withstand temperatures of up to 260 degrees Celsius and has excellent chemical resistance and is therefore used in automotive brake systems. In contrast, most plastics have a broad set of characteristics that enable them to be used across a wide variety of uses. Polyethylene terephthalate, also known as PET, is one such material and is used in the manufacture of drink containers and carrier bags through to the NASA-developed space blanket.

The evolution of plastics from substitute materials to glamour icon and eventually into a high-performance material continues unabated. This trajectory saw plastics first used as a bespoke material, crafted like timber and leather. Today it is more commonly associated with large-scale industry, technological infrastructure and high production quantities. Recent developments in 3D printing or additive manufacturing (AM) have expanded the use of plastics beyond those of the current industrial production paradigm into smaller and more flexible contexts dominated by design and craft studios. The ability for low volume and zero investment allows for smaller operations to produce objects in plastics that once were the sole domain of large industry. This significant shift now locates plastics in a unique position where they are able to be considered for use across the entire production spectrum, from high-tech industry to artisanal craft.

Stage four of plastics evolution saw plastics emerge at the beginning of the twenty-first century able to be manipulated in both large factory contexts as well as small studio workshops. 3D printing is a technology that enables the printing of a physical object in three dimensions. The process requires the creation of a digital model of the part to be printed. The model is virtually sliced into horizontal layers that are sent to a 3D printer. The printer then prints each layer in a physical material one on top of the next and so building a physical 3D representation of the original digital model (Gibson et al. 2014). 3D printing first emerged in 1987 with a process called Stereolithography which solidified thin layers of light-sensitive liquid plastics using a laser (Wohlers and Caffrey 2015, 17, 38). Initially, it was used to print objects for visual evaluation or as appearance models because the materials used did not have enough structural integrity to be functional and the part cost was very high (Markillie 2012). Although 3D printing is now capable of incorporating a wide variety of materials, plastics continues to be the most common material used. Indeed, the emergence and continued growth of 3D printing is due in large part to the characteristic of plastics that as materials they can be easily heated and formed.

The transition of 3D printing from being a tool for prototyping to a process for manufacturing end-use parts has seen the plastics used in 3D printing develop considerably (Markillie 2012, 2). A vast array of materials has been developed for 3D printing that includes acrylonitrile butadiene styrene (ABS), polyanaline (PAL) and polyvinyl alchohol (PVA). The printers have equally evolved from machines capable of printing small buildings in concrete to desktop printers suitable for children. Conventional plastics processing is dominated by large, complex and costly machinery used mostly by highly skilled personnel employed by medium and large companies. This is a significant barrier to micro and small enterprises incorporating plastics into their work practices. It has been suggested that 3D printing is part of a significant shift in manufacturing that is democratising the sector. Where once only large companies produced objects in plastics, now thanks to 3D printing, designers in small studios and artisans in their workshops can as well.

The evolution of traditionally moulded plastics such as injection, extrusion and rotational moulding first mastered perfect aesthetics, that is, the ability to produce homogenous material, high gloss finish and geometric surface precision before adventuring into high-performance material characteristics. Today 3D printing in plastics is at the point where it can produce structurally sound parts equal to injection-moulded ones (Thompson et al. 2016), as long as the parts are designed for 3D printing as the production method. 3D printed parts do not have the smooth surface finish or homogenous material consistency achievable using conventional processing methods (see Figs. 6.1 and 6.2). Despite this, consumers have

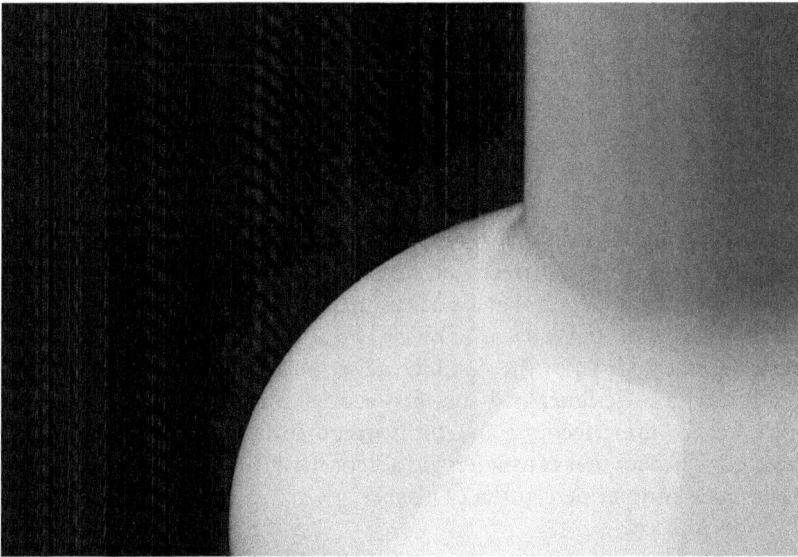

Fig. 6.1   Close-up of traditional plastics product, smooth surface with high gloss finish. Photo: Paul Sutton

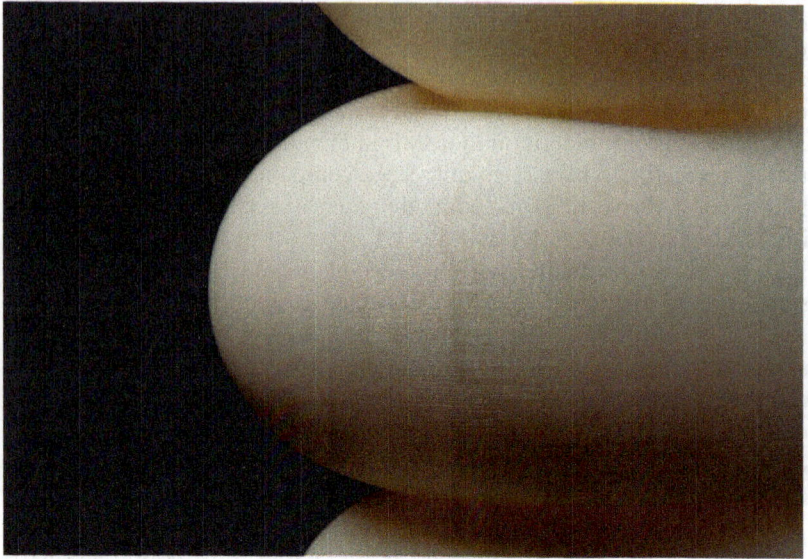

Fig. 6.2 Close-up of a 3D-printed part, showing the layers and slight imperfections. Photo: Paul Sutton

continued to purchase products made by 3D printing. Surveying 3D-printed objects available for purchase at online sites such as Etsy, Shapeways or iMaterialise confirms that objects made by 3D printing typically have rough surfaces or at the very least show evidence of build steps.

More than half (56.3%) of all 3D printing in 2018 was to manufacture end-use parts or functional prototypes (Wohlers et al. 2019, 28). This demonstrates people's growing enthusiasm for designing, manufacturing and consuming products that are 3D printed. And as demonstrated above, it is the designers and artisans of micro and small businesses that are playing a significant role in the commercialisation of 3D-printed outcomes. Previously, if a designer had specified a plastics part that was produced with surface imperfections and form irregularities, it would have been rejected. The fact that consumers are accepting these imperfections signals a shift in consumer perception of plastics.

DISCUSSION

The advance of 3D printing technology continues, despite the apparent flaws in the surface quality and geometric precision. We propose a rationale to explain why 3D-printing technology has continued to advance in high-level, high-precision industrial categories while simultaneously developing as an accessible technology for more general use, despite its apparent limitations. Following Desmet and Hekkert (2007, 59), we consider the three types of product experience that relate to aesthetics, meaning and emotion. As the focus of this chapter is on the aesthetic qualities of plastics, we will incorporate this parameter in terms of instrumental and non-instrumental human-product interactions. Instrumental interactions refer to the understanding of how a product functions including use, operation and management of products. The non-instrumental interactions refer to how one might play with a product or delight in the feel or look of its surface finish. Both aspects are attributable to the aesthetic pleasure of product experience and useful for understanding users' perceptions of the aesthetic qualities of products past and present.

Plastics are highly scientific materials, historically only capable of manipulation through proprietary industrial production processes. Plastics are the product of an industrialised society where the complex circumstances in which products are designed and manufactured have been taken for granted (Mayall 1979) and consequently are poorly understood by the general public. The advent of 3D printing and the subsequent democratisation of the technology signalled a dramatic change in the ability of people to understand plastics and plastics processing more deeply. Although the internet has made information on materials and manufacturing processes more readily available, it is the engagement through making, in this case via 3D printing, that promotes access to knowledge about materials, methods and processes that are otherwise difficult to acquire (Fingleton 1999, 20–21). There are hundreds if not thousands of 3D-printing videos on the internet that make the process explicit, meaning that they show a machine either depositing material or melting it together to build a part. A person watching something like that will, at the same time, both learn how the process works, as well as gain an understanding of why a part produced in this way displays the aesthetic qualities it does. The part now represents a process that is completely understood and can be recounted to others. In contrast to this injection moulding is challenging to understand and explain. No matter how well injection moulding is communicated, the process takes place inside the machine where the actual moulding stage cannot be witnessed.

We now consider the background reasons why 3D-printing technology continues to grow in popularity amongst users, despite the inferior nature of 3D-printed parts over mass-manufactured plastics parts as presented through Norman's reflective human processing (Norman 2004). The reflective level refers to the ability to interpret, understand and apply reasoning. When a person is able to understand the nature and consequences of material and the manufacturing processes for a given product, they are able to understand the product deeply and reflect upon its suitability and purpose in connection with their identity. 3D printing has abruptly opened up this opportunity for plastics in ways that have been, until recently, only accessible to the highly trained designer or engineer. Combining Norman's (2004) reflective human processing level with Hekkert's (2006) aesthetic pleasure, human-product interaction level allows us to hypothesise the course of development of an overarching level of appreciation for plastics across the key evolutionary periods discussed in this chapter. These are:

- Mid-nineteenth century—where plastics were used as a substitute for finer materials.
- Mid-twentieth century—where plastics became highly industrialised and adopted by designers to create non-traditional forms.
- Late twentieth century—where plastics were being developed and utilised in high-performance applications.
- Early twenty-first century—where plastics can be shaped into products by anyone utilising 3D-printing technologies.

Additionally, we must clarify, in the utilisation of Hekkert's and Norman's theories that in framing this rationale, the 'product' is plastics and the 'user' is the fabricator of the object in plastics. We propose general levels of *instrumental aesthetic interaction, non-instrumental aesthetic interaction* and *reflective human processing* between plastics and fabricator based on evidence of the changing purpose of plastics products across four key periods in history, as outlined above.

Mid-Nineteenth Century. *Reflective human processing* of plastics as a material began at a relatively low level. Effort in experimentation was geared towards crafting plastics into forms that replicated the shape and appearance of products made from other (often considered, finer) materials. Because plastics were used as a substitute material and the fabricators intention was to create cheap alternatives, we can infer that the *non-instrumental aesthetic interaction* was also low. Due to the emerging nature of plastics during this period, there was little in-depth exploration of how the specific and unique characteristics of plastics may be translatable into marketable product forms resulting in a low *instrumental aesthetic interaction*. That is to say, that the development of functional attributes that take advantage of plastics unique potential was not foremost and certainly not experienced by end-users, beyond the fact that plastics product 'versions' were easier to access. The exploration and design work necessary to develop and advance instrumental aesthetic qualities was only slowly progressing because of the immature status of the material.

Mid-Twentieth Century. There was a sharp increase in *reflective human processing* of plastics as a material with the formation of the industrial design profession. Industrial designers were able to bring together engineering and marketing teams to develop products for plastics production thereby activating a concurrent increase in the appeal of plastics and advances in industrial technology. As a result of the increased design intervention for the benefit of market penetration, *non-instrumental aesthetic interaction* also increased with a clearer focus on how plastics offered enormous shape and colour variations unlike any other material. Designers, with the support of industry, explored the advantages of the material further pushing the advance of associated manufacturing technologies. The exploration into how plastics' aesthetic qualities could elevate their status as materials consequently increased levels of functionality (*instrumental aesthetic interaction*) which led to a phase of deeper investigation of plastics structural and mechanical advantages.

Late Twentieth Century. A result of the proven advantages of plastics as materials for mass-volume production, coupled with society's acceptance of the materials group encouraged and supported a deeper level of *reflective human processing* in exploring the performance potential of plastics materials on a molecular level. Manufacturers invested heavily in the development of material variants and composites that provided plastics with unique and advanced mechanical properties. Plastics, during this period were widely accepted and understood as the material that certain products were made from, in a sense, defining what those products are. Advances in the materials and their manufacturing processes allowed the products that were made from plastics to have both high *instrumental and non-instrumental aesthetic interaction*.

Early Twenty-First Century. 3D-printing technology addresses two fundamental limitations of plastics as a materials group. The first is the limitation of geometric forms imposed by the industrial method of moulding. Development of plastics as a material and of the sophisticated mass-volume manufacturing techniques to form it into shapes were intrinsically connected together, advancing their commercial success but constraining their ability to adopt any possible form. 3D printing de-couples that connection. The second limitation is access to the material. Despite the vast amounts of plastics in our society, only industrial firms producing large volumes are able to command how the material is used. 3D printing makes the fabrication of plastics parts available to all. All three levels of *reflective human processing, instrumental and non-instrumental aesthetic interaction* relating to both the technical and socio-cultural purpose of plastics have increased dramatically with the advent of 3D printing. The nature of this increase means that the imperfect surface quality of printed parts is not detrimental to the advance of 3D-printing technology nor has it hindered the acceptance of the material.

CONCLUSION

Analysis of the imperfect aesthetic of 3D-printed plastics parts using a combination of Hekkert's *instrumental and non-instrumental aesthetic interaction* and Norman's *reflective human processing* is relevant for two reasons: because plastics are products not just materials and because people value the operational and functional capacity of plastics not just their aesthetics. The first point draws attention to the fact that plastics are man-made materials and, therefore, designed products. It is not simply the material used in the design and manufacture of consumer goods. Consequently, the nature of the interaction between people and plastics is centered around the advance of technologies that enable the formulation and manipulation of plastics materials. This is an important distinction because the literature tends to more directly refer to the way people interact with finished consumer products.

The second point builds on from the first by acknowledging that plastics are designed entities for specifically designed manufacturing systems. Plastics manufacturing systems operate according to particular means by which people interact with the material through conscious engagement, impacting on its aesthetic value. Aesthetic value is the way plastics function and how they are used (*instrumental aesthetic interaction*) together with the way they feel and look (*non-instrumental aesthetic interaction*). When thought about this way, we can say that the *instrumental aesthetic interaction value* between plastics and people has been fulfilled through the material experimentations that plastics have invited throughout history and that is clearly evident in 3D-printing technology. 3D-printing technology is an iteration of the way plastics have, since their creation, invited a highly satisfying level of *instrumental aesthetic interaction*. Essentially meaning that the practices associated with the creation of plastics parts can be, in and of themselves, so highly aesthetically valued that rough or flawed surface finishes on the resultant parts are deemed to be an acceptable trade-off.

To use Hekkert's and Norman's theories as a rationale to embrace the imperfect aesthetic in 3D-printed parts, when it would seem that a sophisticated combination of precision surface and high engineering performance have come to define plastics' unique value, offers the following account:

1. Plastics as a product across the enormous number of forms that they adopt including 3D printing, present opportunities to design into the production and application of the material a balance of both pragmatic and emotive types of engagement (*instrumental and non-instrumental aesthetic interaction*). A characteristic that further reinforces the uniquely flexible nature of the material.
2. Considering the fabricator as the user, in the case of plastics, we can see that historically, the value of understanding of how the product functions (*instrumental aesthetic interaction*) has steadily deepened according to the level of reflective reasoning associated with advancing the manufacturing technologies for the manipulation of plastics. In recent years, 3D printing has become a popular topic of conversation in circles outside of design, engineering and manufacturing. The basic principles of how it works are comparably easy to understand, so more than any other method for the manipulation of plastics, 3D printing has deepened this value of understanding of how products function.
3. We may then argue that the material's surface finish and feel (*non-instrumental aesthetic interaction*) has not been the main driver in defining the material's overall value in our society. Rather that considerably more reflective reasoning by fabricators of plastics materials has been focused on plastics functional characteristics (*instrumental aesthetic interaction*). The huge effort required by

design and engineering to develop technologies to create reliable structures in plastics, a material very difficult to control, associates the meaning of plastics to the means by which they are shaped, far more heavily than how they look. The recent focus on 3D printing and the subsequent broadening of the plastics' user base has contributed significantly to the democratisation of plastics as a material which has led to a more honest (less superficial) appraisal of the material and associated processes.

4. 3D-printing technology is a natural extension of the union between what plastics can do (*instrumental aesthetic interaction*) and the reflective reasoning applied to the development of technologies to shape the material, providing an understanding as to why the imperfect aesthetic is embraced. Just as it has been throughout the history of the material, plastics' surface appearance is not what predominately defines the design of the material, it is the opportunity it provides for plastics to perform new functions and create previously impossible forms. And along with the advantage of geometric freedom offered by 3D printing, many of the internal structural limitations of printed parts can also be overcome (Walden et al. 2020).

This chapter has shown how the aesthetic value in human-object interaction with products made from plastics has significantly changed over time, and how designers are accommodating this change in the design of 3D-printed objects. Since the emergence of man-made plastics materials in the mid- to late nineteenthth century to the recent proliferation of plastics including 3D printing, the understanding by people of the material has varied. Society's perception of plastics has evolved from a material prized for what it represented to a material prized for its performance, and then to a material that is prized for its ability to allow for unlimited design freedom.

Design freedom and the democratisation of plastics' processing is enabling a broader range of engagement with plastics from across the spectrum of society, not just from designers and engineers, which is resulting in a more widespread understanding and appreciation of the materials group. Although the ubiquitous nature of plastics and the resultant impact on the environment is proving to become one of humanities biggest challenges, comfort should be taken from the new aesthetic value people place on plastics as a way to ensure it is appropriately valued and effectively utilised.

BIBLIOGRAPHY

Annicchiarico, Silvana, ed. 2001. *1945–2000 Design in Italy*. Ed. La Triennale di Milano. Rome: Gangemi Editore.

Bosoni, Giampiero. 2008. *Italian Design*. New York: MoMA (The Museum of Modern Art).

Catterall, Claire. 1990. Perceptions of Plastics in Britain, 1945–1956. In *The Plastics Age: From Modernity to Post-Modernity*, ed. Penny Sparke, 66–73. London: V&A Publications.

Desmet, Pieter, and Paul Hekkert. 2007. Framework of Product Experience. *International Journal of Design* 1 (1): 57–66.

Fenichell, Stephen. 1996. *Plastic: The Making of a Synthetic Century*. New York: Harper Business.

Fingleton, Eamonn. 1999. *In Praise of Hard Industries: Why Manufacturing, Not the Information Economy, Is the key to Future*. Boston: Houghton Mifflin Company.

Gibson, Ian, David Rosen, and Brent Stucker. 2014. *Additive Manufacturing Technologies: 3D Printing, Rapid Prototyping, and Direct Digital Manufacturing*. 2nd ed. New York: Springer.

Hekkert, Paul. 2006. Design Aesthetics: Principles of Pleasure in Design. *Psychology Science* 48 (2): 157–172.

Kamitsis, Lydia, and Paco Rabanne. 1996. *Paco Rabanne, les Sens de la Recherche = Paco Rabanne: A Feeling for Research*. Paris: Editions M. Lafon.

Kaufman, Morris. 1963. *The First Century of Plastics: Celluloid and Its sequel*. London: Plastics Institute, Distributed by Iliffe Books.

Manzini, Ezio. 1990. And of Plastics. In *The Plastics Age: From Modernity to Post-Modernity*, ed. Penny Sparke, 132–143. London: V&A Publications.

Markillie, Paul. 2012. Special Report—Manufacturing Innovation: Solid Print. *The Economist*, 8–14.

Mayall, W.H., 1979. Principles in design. Design Council.

Meikle, Jeffrey L. 1995. *American Plastic: A Cultural History*. Brunswick, NJ: Rutgers University Press.

Mossman, Susan. 2008. *Fantastic Plastic: Product Design + Consumer Culture*. London: Black Dog Publishing.

Norman, Donald A. 2004. *Emotional Design: Why We Love (or Hate) Everyday Things*. New York: Basic Books.

Sparke, Penny. 1990. On the Meanings of Plastics in the Twentieth-Century. In *The Plastics Age: From Modernity to Post-Modernity*, ed. Penny Sparke, 76–13. London: V&A Publications.

———. 1999. *A Century of Design: Design Pioneers of the 20th Century*. London: Mitchell Beazley, an imprint of Octopus Publishing.

Spivack, Emily. 2012. Stocking Series, Part 1: Wartime Rationing and Nylon Riots. *Smithsonian Magazine*. https://www.smithsonianmag.com/arts-culture/stocking-series-part-1-wartime-rationing-and-nylon-riots-25391066/. Accessed April 15 2020.

Thompson, Mary Kathryn, Giovanni Moroni, Tom Vaneker, Georges Fadeld, R. Ian Campbell, Ian Gibson, Alain Bernard, et al. 2016. Design for Additive Manufacturing: Trends, Opportunities, Considerations, and Constraints. CIRP Annals: Manufacturing. *Technology* 65 (2): 737–760. https://doi.org/10.1016/j.cirp.2016.05.004.

Walden, Roderick, Stefan Lie, Berto Pandolfo, and Anton Nemme. 2020. Developing Strategic Leadership and Innovation Capability for Manufacturing SMEs Transitioning to Digital Manufacturing Technology. In *Leadership Styles, Innovation, and Social Entrepreneurship in the Era of Digitalization*, 164–189. IGI Global. https://doi.org/10.4018/978-1-7998-1108-4.ch007.

Wohlers, Terry, and Tim Caffrey. 2015. *Wohlers Report 2015: 3D Printing and Additive Manufacturing State of the Industry Annual Worldwide Progress Report*. Fort Collins, CO: Wohlers Associates Inc.

Wohlers, Terry, Ian Campbell, Olaf Diegel, Ray Huff, and Joseph Kowen. 2019. *Wohlers Report 2019: 3D Printing and Additive Manufacturing State of the Industry*. Fort Collins, CO: Wohlers Associates Inc.

CHAPTER 7

Witches' Knickers and Carrier Bag Theories: Thinking Through Plastics

Joanne Lee

Plastics are frequently characterised in a reductive binary: *either* they are beautiful and adaptive *or* they are unauthentic, destructive and polluting. In this chapter, I want to suspend such definitive value judgements a while, and instead sustain a more complex and poetic critical encounter with the material. I have developed this approach through my artistic-research for an independent artist's publication series in which I use everyday materials and phenomena as matter to think with and through: previously I have considered littered chewing gum, the graffiti that proliferates on urban walls and the litter that ends up in the vaguer terrains of urban edgelands. In thinking through forms, materialities and meanings, I seek to develop a richer relationship with seemingly ordinary stuff. Whilst this is intended to deepen my engagement with the complexities of the everyday, the matter with which I engage also starts to rethink what the research is doing and begins metacritically to shift its possibilities.

This chapter draws on artistic research for 'Witches' Knickers', an edition of my serial Pam Flett Press, named after the coinage for carelessly discarded plastic bags caught on trees and fences. (The term is thought to

J. Lee (✉)
Sheffield Hallam University, Sheffield, UK
e-mail: joanne.lee@shu.ac.uk

have emerged in Ireland around 2000 and since then the term has come into more common usage; by 2015, it even found itself the subject of a 2015 episode of Radio 4's *Word of Mouth* (2015)). I explore the ways in which plastics (particularly in the form of bags) feature in art practice, music, film, material culture and waste studies. As well as focusing on the specific problems and possibilities of plastic bags, I develop upon Ursula K. Le Guin's idea of 'the carrier bag theory of fiction' in which books are bags for holding words and ideas 'in particular, powerful relation to one another and to us' (Le Guin 1989, 169). The bag, and the artist's publication through which I research, is thought of literally and metaphorically as a handy container for holding together differently valued, and potentially contradictory ideas about our use of and relationship with plastics.

SITE AND MATERIALITY: ARTISTS AND WRITERS INVESTIGATE PLASTIC BAGS

As I have investigated plastic bags, I have realised I am far from alone in my artistic fascination for this humble object: a host of creative projects have explored the topic. There was, for example, Zoe Leonard's *tree + bag* photographic series, taken over several years in New York. In an interview, she explained how she 'kept noticing a particular place where the wind catches stray plastic bags and floats them up into the branches of these two small trees. After a windy day there could be four or five, or as many as a dozen, plastic bags hanging off the branches' (quoted in Dungan 2002, 70). Leonard said how she loved 'the beauty and ugliness of it. It's an odd image: cheerful and depressing at the same time' (70). In one black-and-white picture, a group of residents sit on benches beneath the spindly trees: a collection of shopping bags at their feet mirror the empty bags suspended in the branches above. In another work from the series, four dye transfer colour prints present the same trees shown on different occasions, the passage of time denoted by the absence or presence of lingering snow, and with varying amounts of bags caught there. In this latter example, the long 'season' of the bags, sustaining there till they are blown free or shredded by winter storms (or removed by civic authorities) has something to reveal about the duration of our anthropogenic effect upon the world.

It was a phenomenon also noticed by someone else resident in New York at the time. Writer James Frazier opened the first of three essays on the subject with the lines: 'This is the season of plastic bags stuck in trees. Stray shopping bags—many of them white, with handles, perhaps from a deli or a fruit-and-vegetable store originally—roll along the streets, fill with air, levitate like disembodied undershirts, fly, snag by their handles in the branches' (Frazier 2006, 61). This piece from 1993, simply entitled *Bags in Trees*, went on to itemize the different examples that he encountered in Brooklyn: 'I saw yellow Tower Records bags, tan bags, red bags like the kind you get in Chinatown, Key Food bags, C-Town bags' (61). He remarked on the way they change: 'Last season's bags have shredded into a sort of plastic Spanish moss. Big black garbage bags become entangled in branches in a sprawling and complicated way' and how over time 'light-colored bags darken and dark ones fade until all are a variety of gray' (62). He noted too how the bags collect ultrafine traffic soot that comes off on his hands like graphite. A year on, in the second article, he was so affected by the number of bags in trees that together with a friend he devised an implement with which to untangle them; with it they are able to reach high into branches and pluck free the material that had been caught there. In the final instalment of 2004, he told how he and his friend used their invention, regularly heading out for a few hours 'bag snagging', an activity that had for them the thrill of vandalism but that was 'mischievous good rather than mischievous bad' (154). They went on to patent this device, even setting up a not very successful company to market and produce it in volume. What is most interesting to me from the writing is the way that attempting to remove bags made Frazier very attentive to the diverse materiality of the plastics he encountered. He described the filmy whiteness of New York deli bags and noted how easy they were to remove when fresh, and how impossible when aged. He also saw how the very thinness of 'the micron-thick plastic of a dry-cleaning bag' when shredded by wind for only a day or two made them 'all but unremovable' (157). And ultimately he ascribed a ghostly agency to the type of white deli bag that he saw most frequently, suggesting that 'It does not have a soul but it imitates one': he reckoned that these bags seemed to have got into the branches on their own initiative, and even that 'They like it up there and prefer not to be disturbed' (157).

In Manchester, UK, the artist Hilary Jack had also noticed a bag flapping in the branches of a tree near her studio: the blue specimen seemed so startling and vibrant against the colourless cityscape that it prompted a new project, that of the *Turquoise Bag in a Tree*. Once alert to the phenomenon, she started to see them everywhere and went on to photograph many more instances in what became a series commissioned by Cornerhouse, Manchester. Eventually, she presented them on a website dedicated to turquoise bags in trees everywhere. The bags there are mainly revealed by leafless winter trees; for the most part they occur amidst utilitarian housing estates, public footpaths and adjacent to vacant urban spaces, though one of her images shows a turquoise flash against the silver bark of birches artfully planted outside the Tate Modern gallery in London. Writing on the site in 2009, some six years after the project's inception, she said the original bag 'still clings on' although 'torn and dishevelled'; she described it as 'an unsettling and melancholic memorial to our excesses', reflecting on the way that such matter will seemingly long outlive us (Jack 2009).

In South East London, photographer Rosie Barnes (n.d.) was also seeing bags caught in trees, and drifting or deposited in the landscape. Her series Witches' Knickers was beautifully shot on analogue film. In one image, a thin pinkish-red and white striped specimen, of the type familiar from off licences and corner shops, is caught on a frond of conifer, such that it appears to hover between dark trees, and in another it is there suspended upside down amongst the tips of elegant long grass. In a further photograph, she has an orange example: its colour immediately suggests that it must likely have originated from a Sainsbury's supermarket and I realise in identifying brands in this way how as consumers we ably mirror the skills of naturalists or birders in spotting particular species. The orange bag is echoed by a graffitied circle of a similar colour on a brick wall behind, and one elsewhere picks up the autumn colours of a Virginia creeper; they seem like those octopuses who mimic their habitat. In such examples, the bags are almost creaturely; they are artificial beings existing amidst the parks, on streets and in everyday places.

Whilst the pictures appear to have something theatrical and staged about them, the scenes were in fact simply as she encountered them: other than the formal decisions of how to frame the scene and in what light conditions to photograph, the images were not manipulated in subsequent post-production. The pictures did not campaign explicitly about the issue of littered bags: whilst Barnes was concerned about the issue, she was keen to prevent people switching off or feeling inured to the problem. She said: 'With so many "news" stories we're so accustomed to seeing "doom and gloom" that it almost stops touching us. So, I try to make a visually stimulating image that will stay with you and make you think about the subject, perhaps in a different way' (quoted in Sevier 2009). This sense of staying with the phenomenon, and not being too quick to offer judgement, recurs in many of the examples of artistic practice I have considered. There is a fascination, a curiosity about the object and the fact of their appearance in our daily lives that is perhaps best described as a kind of care, something that I will return to later.

Countless photographers have been fascinated by the Witches' Knickers. Users of social media sites such as Flickr have frequently attended to snagged plastics: 'Ian@NZFlickr' shows a tattered black sheet (perhaps from a bin liner, though maybe it is a fragment of the material used to wrap big silage bales in agriculture) snagged on a barbed wire fence, set against a misty landscape of rough pastureland, single storey buildings and trees, whilst 'mad jeff' has white stuff snared by a serpentine of vicious-looking razor wire. James Frazier said that there is little scarier than 'bare-branch trees draped with plastic-bag shreds above a razor-wire fence similarly fluttering and bestrewn' (157). The Instagram pictures posted by Hull-based Kevin Rudeforth reveal a veritable obsession with the stuff: in one a billow of fraying white plastics flares from a razor-wired security fence; in another the material is shredded by wind to form silvery tresses; and elsewhere it seems transformed by the camera into the image of a frozen water splash.

This attention to the sensuousness of plastics, their thickness or trans-lucency, their folds, pleats and creases, rather akin to that paid to drapery in Baroque art, was also manifest in the work of contemporary painter Marguerite Heywood. Her pictures of discarded industrial plastics in and around a Bolton industrial estate revel in their surfaces and textures, in their reflections and the absorption of light as it lingers in shadowy corners or floats against grey northern skies. Heywood considered her paintings a form of still life, a genre that traditionally references ideas of mortality: by working with plastics, she has said she is dealing instead with 'an unimag-inably long decline rather than the usual quick death' (quoted in Monk 2014, 16). She too holds the positive and negative in tension, recognising the beauty of plastics, and that they are a technological triumph, but also that they are a terrible environmental problem. Ultimately, she has said that her intention is to make people really look at plastics and 'perhaps ponder its place in our world' (15).

THE AFFECTIVE AND ETHICAL DIMENSIONS
OF PLASTIC BAGS

This sort of artistic engagement as a way of considering bags and their place in the world brings me to Gay Hawkins' chapter on the subject in *The Ethics of Waste: How We Relate to Rubbish* (2006). Hawkins reflected on her response to the film-within-a-film, the so-called 'rubbish scene', in Sam Mendes' *American Beauty*, when a discarded carrier bag is shown swirling lyrically, sensuously, *enchanted* in an eddy of wind. She noted how as she watched she felt haunted by the whole, enormous weight of eco-logical crisis: whilst she recognised how the scene attempts to transform the worthless and trivial into the beautiful, because there seemed no con-cern for the moral or ecological consequences, at first she felt far too ambivalent to accept its apparent redemption (Hawkins 21).

But the invitation to change her feelings *does* begin to have an effect: in being encouraged to find delight in something she had been trained to hate, she started to feel a strange sense of sympathy. This reminded her of 'the complex social life of bags', and that before they clog up our kitchen drawers and make us start to feel guilty about their eventual fate as landfill, they are incredibly handy containers: after all, there are so many occasions when we find ourselves gratefully saying 'oh, here, put it in a plastic bag' (23). Hawkins pointed out the 'very ambiguity and complexity of uses that complicates moral rulings about and condemnations of bags as bad' (23). In feeling sympathy, she experienced a sincere kind of ethical con-cern; indeed, she claimed that the 'rubbish scene' is much more effective than an advert from the New South Wales Environmental Protection Agency designed to stop people discarding bags, at which she had felt guilty, patronised, and irritated by its moralism and pedagogical intent. Her point was not to defend plastic bags but to acknowledge the variety of relations we have with them. Hawkins suggested that the film locates plastic bags 'in the realm of the sensual and the affective' (22) and as a result 'it refuses the essentializing move that renders rubbish always already bad, denying paradox and ambiguity—let alone any recognition of our shifting relational sensibilities with it' (22–23).

Hawkins drew upon Michel Foucault's *The Use of Pleasure* (1985) to consider the arts of existence—'all those actions and rules of conduct through which we organize ourselves according to particular ethical and aesthetic criteria' (24)—as a means of considering how we dispose of our waste, and how that disposal might be 'recast in ways that acknowledge the ethical significance of rubbish without generating moral righteousness or resentment' (36). And she turned to Gilles Deleuze and Felix Guattari's *A Thousand Plateaus* (1987) for 'an ethics that has little to do with rule-bound moralities', considering that their 'insistence on the plurivocity of being, on the dimensions, lines, and directions, signals how movements of becoming and immanence fragment the normative work of macropolitics' (36). Finally, she drew on William Connolly's argument in *Why I Am Not a Secularist* (1999) that 'conscience and other code-driven moral techniques are crude and blunt tools for coping with the world', recognising that such moral techniques tend to suppress 'the visceral and situational' (37). In our encounters with waste, of course, 'the visceral is ever present and palpable': whilst we may be acting out of concern for the environment, out of duty and guilt, we are also encountering dirty, broken, rotting stuff which has lost its use and value, and that has a powerful physical and emotional effect (37). Hawkins suggested that our responses to this material are 'shifting and surprising', and that these are moments of intensity—as perhaps when one encounters a fecund mass of worms making compost, or when we recoil with aversion at rubbish spilled on the road after collection day. She noted how these moments 'surprise and unsettle' and linked this to Connolly's sense that 'responsiveness is a condition of possibility, it opens up lines of mobility and difference within the self, and it is something that can be cultivated' (38).

She concluded with a reference to an essay by Italo Calvino on the household ritual of emptying the bin, describing how it brings together thinking and feeling, the spiritual with the intellectual, through his generosity to experiencing waste. For Calvino, the gesture of throwing away was the 'first and indispensable condition of being' and this disposal an offering 'made to the underworld, to the gods of death and loss' (Calvino 1993, 104). Ultimately, as Hawkins read Calvino and thought of the rubbish scene in *American Beauty*, she identified the 'messiness and ambiguity that makes ethical work experimental, creative, and relational' (41). Italo Calvino and Sam Mendes' work generated, she claimed, a much wider range of responses than mere moralism, and thus they are ultimately more effective in ethical terms. These responses and the opening up of lines of mobility and difference are, I think, generated in the writing, photographs and paintings I have previously described and in similar poetic 'rubbish scenes' from Jem Cohen's *Lost Book Found* (1996) and in Ramin Bahrani's *Plastic Bag* (2010) (in the latter, Werner Herzog narrates an anthropomorphised plastic bag's 'life' as it 'journeys' through the world.)

A willingness to pay attention to the usually overlooked has clearly been a feature of recent creative practice and in many cases there is a clear desire to suspend judgement such that transformative aesthetic and therefore ethical work can be done. Kevin Newark's photographic series *Protoplasm*, which pictured discarded bags floating in the Regent's Canal in London, is another case in point (Newark 2009, *Protoplasm*). Photographed from above, Newark reveals their transparent, white or blue plastic forms floating against inky water that is speckled with other tiny pieces of debris. We might rightly feel disgust or anger at the littered environment, but in Newark's images the bags also appear as the beautiful cellular forms suggested by his title, and even, since the scale is carefully confused, as if we are encountering some sort of galactic/astronomical scene reminiscent of the pictures one might see relayed by NASA; as a result, we are encouraged to stay with them a while, just looking, and thus to feel a wider range of emotional, visceral, poetic and intellectual responses.

But this complexity is present not just in artworks and film, it is there too in more 'everyday' situations with bags: I think here of a shocking image I found on a global travel site which shows an Acacia tree in Niger with few leaves but very many windblown black bags. Site user niamey00 tells how plastic bags became caught up in its thorns and were then extremely hard to remove; as a result, the tree became a landmark in the neighborhood and was used when giving directions: 'When you come to the tree with all the plastic bags in it, turn left' (niamey00 2004). Here, a negative situation is put to a kind of 'useful' work.

I have been trying to attend to this complexity in my own work on bags. In one image from a photographic series, I saw how at home were the snails on discarded plastics that others might think of as despoiling the landscape, the damp folds forming a safe retreat (Fig. 7.1). The molluscs' presence there reminded me of artist Rebecca Beinart's comment during *Desire lines: Art, Edgelands and the Urban Wilds*, a symposium organised by Manchester commissioning agency Buddleia: she told how an ecologist had explained that the much-maligned abandoned shopping trollies tossed into urban waterways can in fact serve as effective nurseries for small fish, as well as providing a useful anchorage from which plants can grow. The apparently negative is obviously not wholly so.

Plastics waste does, of course, have very problematic ecological consequences, as was made evident to television viewers in the final episode of the BBC Blue Planet II series, where filmmakers told of its pervasive and devastating effects in the marine environment. Those watching were

Fig. 7.1 *Snails* by Joanne Lee from the photographic series 'Witches' Knickers', 2015. Commissioned by Harris Museum and Art Gallery, Preston

shocked by the footage of entangled and poisoned wildlife, many vowing on social media to renounce or reduce plastics in their daily lives as a result. However, it is not just the large scale, very visible material that is dangerous, but the volume of small and microscopic pieces of plastics floating in our seas and waterways. According to the results of a six-year study by the 5 Gyres Institute, an estimate was made that 5.25 trillion plastics particles weighing around 269,000 tons are floating on the surface of the sea (Eriksen et al. 2014). Researchers such as ecotoxicologist Heather Leslie of VU University Amsterdam have been exploring how the particles of plastics entering the food chain through being consumed by fish and other aquatic species and then by humans 'can induce immuno-toxicological responses, alter gene expression, and cause cell death'. Carol Kwiatkowski of the Endocrine Disruption Exchange has considered how our consumption of this material may cause hormonal disruption (Seltenrich 2015, 41). That imaginative work had in some sense already addressed this situation years before as revealed in a comment from Norman Mailer: 'I sometimes think there is a malign force loose in the universe that is the social equivalent of cancer, and it's plastic. It infiltrates everything. It's metastasis. It gets into every single pore of productive life. I mean there won't be anything that isn't made of plastic before long. They'll be paving the roads with plastic before they're done. Our bodies, our skeletons, will be replaced with plastic' (Mailer 1988, 321). In one aspect of my 'Witches' Knickers' edition of the *Pam Flett Press*, I find myself working creatively via instances of speculative fiction to imagine just what this growing 'plastification' of humans and the planet might feel like.

AESTHETICS AND ACTIVISM: WHAT PLASTICS BAGS DO

I have also been exploring the aesthetic qualities of the seemingly visually unredeemable, as for instance, in the case of those plastic bags used to pick up dog faeces, but which their owners then toss aside or hang from trees and fences rather than placing in the bin as they ought. Whilst I still feel the frisson of anger at their failure to dispose of the bagged waste, I have become fascinated by its occurrence: in one photograph I made, a reddish-pink and white-striped bag is depicted with a tightly knotted fastening at the jaunty angle reminiscent of a comedy bow-tie, its colour matching the mallow flower in an adjacent image; in another, a white bag dangles casually from the branch onto which it has been slung such that it has an elegant insouciance. Of this latter practice, one friend of mine suggested that they might even be called 'poo pompoms' or 'doggy baubles': I note that the terms offer a similar sense of decoration and despoliation to that encountered in Zoe Leonard's *tree + bag* series (Fig. 7.2).

Others too have attended to the bagged poo: John Darwell's *DDSBs*, a series of forty photographs made into a self-published book, offer a typology of the form. Julia García Hernández has said it shows Darwell's interest in the patterns of human behaviour 'that at first appear eccentric but which form part of the bigger picture of human impact on the environment and our natural world' (Garcia Hernadez 2013, 17). It strikes me that the strangeness of our ordinary habits and practices is surely present even in the most conscientious collection and disposal of dog waste in plastic bags. An article by Megan Lane on the BBC website pondered: 'What snapshot of our age will future archaeologists deduce when they unearth centuries-old bags of dog poo?' (Lane 2007) (and I imagine those future humans excavating our landfills, wondering about the curious

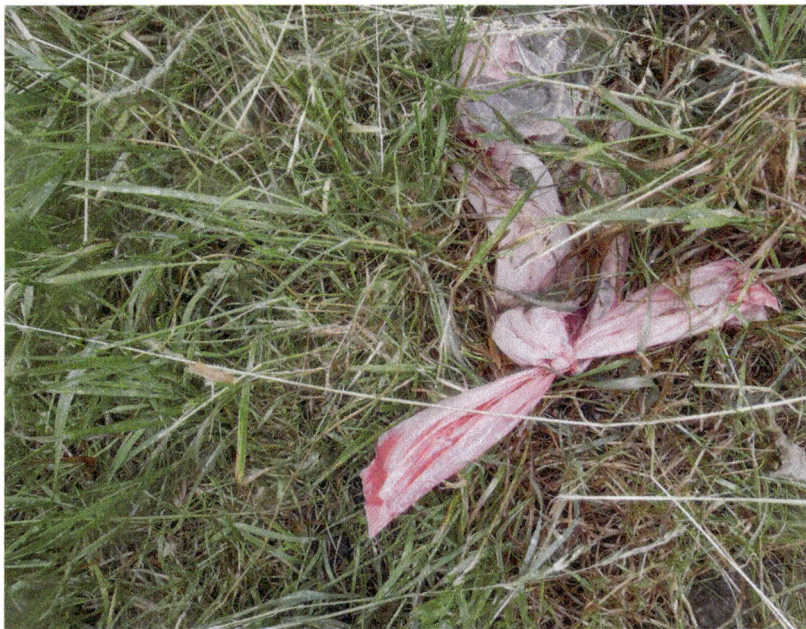

Fig. 7.2 *Knotted* by Joanne Lee from the photographic series 'Witches' Knickers', 2015. Commissioned by Harris Museum and Art Gallery, Preston

meaning of this ritualized bagging of dog waste, and why it was buried with so many of our worldly possessions.

What is currently so ubiquitous, so unworthy of scrutiny, will surely take on interpretative significance as years pass and people seek to understand previous cultures. In a modest way, this is already notable from James Frazier's 1993 essay: the Tower Records bag he mentioned is now an historical artefact of sorts, since the stores are long closed. He also reckoned at the time that close behind the volume of plastic bags in trees were 'furlongs' of audiocassette tape and videotape: now that such formats have been superseded by the CD and DVD, and more recently, by the advent of streaming services, such material is very unusual to find as litter.

Such examples have generated concepts and content for the 'Witches' Knickers' edition of the *Pam Flett Press* as I seek to create a critical shimmer of affective response and to explore the 'messiness and ambiguity that makes ethical work experimental, creative, and relational' (Hawkins 2006,

41). I have also looked to a range of instances where bags are used and visualised in activist practices, in popular culture and fashion. Those protesting littering effects have used vast numbers of the bags themselves to create frightening characters of various sorts with which they have taken to the street in actions designed to attract attention and change behavior. Sometimes these were figured as monstrous and haunting, as when they formed the basis of campaigns such as Andy Keller's 'Don't Feed The BAG MONSTER' (Keller 2020) and that produced by Shanghai's Roots and Shoots '*Waste plastic bags and they'll come back to haunt you*' (Shanghai Roots and Shoots 2011). But what was clearly meant to be frightening in these instances becomes playful in other hands: Joshua Allen Harris, for example, used plastics taped together to create his delightful *Air Bear* character. Positioned carefully over the grates of New York's subway system, when trains approached and departed they were inflated with warm air and rose, quivered and fell as if momentarily alive (Harris 2008).

I was interested to discover that volunteers seeking to provide practical support for the homeless have knitted yarn made from shredded plastic bags—so-called 'plarn'—into mats and cushions to make sleeping rough a little more comfortable (Plastic Bag Sleeping Mats for the Homeless 2009). For some living on the streets, plastic bags are essential tools to enable the carrying of one's worldly goods, and when 'worn' over shoes or clothes can provide some protection against poor weather. That German *Vogue* saw fit to present their version of a 'bag lady look' in Signs of the Times, a spread for the October 2012 edition, where expensive garments were teamed with plastic carriers worn or carried by an elegant model, seemed insensitive and ethically dubious at the very least, but perhaps this photo shoot does reveal something of the complexity Hawkins identifies (Signs of the Times 2012). It is there too in the mixed messages about consumption of the famous 'I'm not a plastic bag' campaign produced for Sainsbury's by fashion designer Anya Hindmarch and the We Are What We Do non-profit-organisation (Hindmarch 2007). Alongside its awareness-raising project, which encouraged discussion about the role of plastic bags in daily life and made clear that people could make choices about their use, the cotton tote became a must-have fashion item in its own right with 80,000 people queuing for one on the day of its launch. It was something desirable to be seen with, or to be squirrelled away as a collectable, to be traded subsequently on eBay, where I note examples in good condition are still selling for £20 or more and where collectors are concerned to spot and avoid fakes.

GATHERING MATERIAL AND GENERATING
MULTIPLE PERSPECTIVES

As Gay Hawkins suggested many of our homes probably have cupboards of the things. A repository of bags is of course incredibly handy to have ready to contain all sorts of messy business: as a keen allotmenteer, I do not know what I would do without them for transporting muddy leeks from my plot. I am fascinated by the ubiquity of these bag hoards, so much a part of everyday life that 'A plastic bag full of other plastic bags' topped a Buzzfeed list of 39 Weird Things You'll Find In Every British Family Home (Finnis 2015). Since the levy on bags has come into force, these collections will doubtless diminish and maybe disappear. These days the remaining bags we have from supermarkets, corner shops or off-licences are likely stuffed in alongside 'bags for life' and the cotton or jute totes that have become the symbol of green credentials, even though in some cases the alternatives may be more resource-heavy or polluting in their production than their counterparts made from plastics. Some people—myself included—even have a few 'special' plastic bags that they keep for sentimental reasons, perhaps because they came from a particular shop or purchase which had personal meaning, or maybe because the bag can be touted for its designer or aspirational credentials.

Such hoarding draws to mind a 1988 track by British band The Fall entitled 'Carry Bag Man', which also formed part of the 'Witches' Knickers' research. Its lyrics refer to 'Carrier bags strewn all around the room' (an image now very familiar from the countless popular TV programmes attending to those whose domestic hoarding has got out of control) and describe how even though the protagonist, the carrier bag man himself, does not sit down, 'I still need armchairs round my home / To put carrier bags on'. Maybe we all know a carrier bag man? In my experience, this is an older chap with a 'bag for life' whose handles are stretched, etiolated, the faded bag stuffed with items picked up in charity shops or found on the street, things considered valueless by many others. The Fall's cantankerous front man, the late Mark E. Smith, was rarely seen without a plastic bag. Interviewers always mention it, walking through the cities of Salford and Manchester, and often on stage with a bagful of lyrics, cans of lager and goodness only knows what else besides. I am interested in Mark E. Smith's embrace of the plastic bag and its relationship to the content of his lyrics, which in some cases is material gleaned from the close observation of the eccentricities of daily life, and the peculiarities of television/

popular culture. Here the bag is both practical prop and creative gathering tool. Zoe Mendelson's *This Mess Is a Place* project made clear how hoarding is considered a pathology in everyday life, but that for artists a propensity for collecting and a studio full of stuff is positively encouraged (Mendelson 2013). She explored hoarding, collecting and sorting, from the layered accumulations of Francis Bacon's studio to the installation practice of Song Dong, whose *Waste Not* installation exhibited some 10,000 items (including a set of carefully folded shopping bags) collected by his mother Xiangyuan across five decades of her life in China.

I want to develop the importance of such activities of collection in relation to the role of the bag itself. In Elizabeth Fisher's feminist conception of the carrier bag theory of human evolution, she agreed with other theorists that 'the earliest cultural inventions must have been a container to hold gathered products and some kind of sling or net carrier' (Fisher 1980, 56). Writer and critic Ursula Le Guin drew on this to articulate the type of novel she was trying to write; she opposed the masculinist, heroic story told through the use of early tools such as bones, arrows, spears, and knives and extrapolated instead the way bags can gather multiple and very different things allowing different stories to be voiced. She said, 'I would go so far as to say that the natural, proper, fitting shape of the novel might be that of a sack, a bag. A book holds words. Words hold things. They bear meanings. A novel is a medicine bundle holding things in particular, powerful relation to one another and to us' (Le Guin 1989, 169).

I am interested in using things to think about and with. As I work to develop my 'Witches' Knickers' edition, the plastic bag is becoming both the subject matter under investigation and a creative critical metaphor for the process of my research and its articulation. Just as all sorts of items can be put into a plastic bag in real life (one's sandwiches, an iPad, a work of philosophy, a child's toy) using the bag conceptually offers a way of bringing together apparently disparate material, and of holding differently valued and potentially contradictory ideas in tension. I have realised that I am applying a carrier bag theory to my research, one which is not about making a definitive, conclusive point and being narrowly incisive (the metaphors of cutting tools in Fisher/Le Guin's proposition), but is rather about using the capacious carrier bag to accommodate and articulate multiple perspectives. This is a type of critical activity closer to the way that some artworks or novels function than to that of much academic writing, and it seeks to expand methodological possibility. In order to do justice to the rich complexities of plastic bags, I need to recognise their actual and conceptual multivalency; like Hawkins I am not making a plea, trying to hymn their positives and elide their more problematic aspects, but rather I try to explore and amplify the very *plasticity* of their uses and meanings.

BIBLIOGRAPHY

Bags Stuck in a Tree? No, They're Witches' Knickers! 2015. *Word of Mouth*, BBC Radio 4, 7 April.

Barnes, Rosie. n.d. Rosie Barnes Photography. *Witches' Knickers*. http://www.rosiebarnes.com/content/32154/projects/witches_knickers. Accessed 7 November 2019.

Calvino, Italo. 1993. La Poubelle Agréée. In *The Road to San Giovanni*, 99–126. London: Jonathan Cape.

Connolly, William. 1999. *Why I Am Not a Secularist*. Minneapolis: University of Minnesota.

Deleuze, Gilles, and Felix Guattari. 1987. *A Thousand Plateaus: Capitalism and Schizophrenia*. Trans. B. Massumi. Minneapolis: University of Minnesota Press.

Dungan, Beth. 2002. An Interview with Zoe Leonard. *Discourse* 24: 70–85.

Eriksen, M., et al. 2014. Plastic Pollution in the World's Oceans: More than 5 Trillion Plastic Pieces Weighing over 250,000 Tons Afloat at Sea. *PLoS ONE*. https://doi.org/10.1371/journal.pone.0111913.

Finnis, Alex. 2015. 39 Weird Things You'll Find in Every British Family Home. *BuzzFeed*. https://www.buzzfeed.com/alexfinnis/the-draw. Accessed 1 October 2015.

Fisher, Elizabeth. 1980. *Woman's Creation: Sexual Evolution and the Shaping of Society*. New York: McGraw Hill.

Foucault, Michel. 1985. *The Use of Pleasure*. Trans. R. Hurley. New York: Vintage.

Frazier, James. 2006. *Gone to New York*. London: Granta.

Garcia Hernadez, Julia. 2013. DDSBs (Discarded Dog Shit Bags) by John Darwell. *Royal Photographic Society Contemporary Photography Journal* 53: 16–18.

Harris, Joshua Allen. 2008. Air Bear. https://youtu.be/L-a607j2dOo. Accessed 12 August 2015.

Hawkins, Gay. 2006. *The Ethics of Waste: How We Relate to Rubbish*. Lanham: Rowman and Littlefield.

Hindmarch, Anya. 2007. I'm Not a Plastic Bag. https://us.anyahindmarch.com/en-US/not-a-plastic-bag.html Accessed 3 May 2020.

Jack, Hilary. 2009. Turquoise Bag in a Tree. http://www.turquoisebaginatree.co.uk/page2.htm. Accessed 12 August 2016.

Keller, Andy. 2020. Bag Monster. http://www.bagmonster.com/. Accessed 5 May 2020

Lane, Megan. 2007. In Defence of Plastic Bags. *BBC News* http://news.bbc.co.
uk/1/mobile/magazine/7071182.stm. Accessed 15 August 2015.

Le Guin, Ursula. 1989. *Dancing at the Edge of the World.* London: Victor Gollancz.

Lost Book Found. 1996. Film. Jem Cohen: United States.

Mailer, Norman. 1988. *Conversations with Norman Mailer.* Ed. J. Michael
Lennon. Jackson: University of Mississippi Press.

Mendelson, Zoe. 2013. *This Mess is a Place.* https://thismessisaplace.co.uk/.
Accessed 12 August 2015.

Monk, Christopher. 2014. All Present and Skip Scape. *The Big Issue in the North,*
25–31 June.

Newark, Kevin. 2009. *Protoplasm. Photographic Exhibition.* Leeds: Pavilion.

niamey00. 2004. *Garbage.* Virtual Tourist. http://members.virtualtourist.com/
m/p/m/d9e5b/. Accessed 1 August 2015.

Plastic Bag. 2010. Film. United States: Ramin Bahrani.

Plastic Bag Sleeping Mats for the Homeless. 2009. https://youtu.be/
BiUC0iDQtkA. Accessed 14 August 2015.

Seltenrich, N. 2015. New Link in the Food Chain? Marine Plastic Pollution and
Seafood Safety. *Environmental Health Perspectives* 123: 34–41. https://doi.
org/10.1289/ehp.123-A34.

Sevier, Laura. 2009. Plastic Bag Art: How to Spoil any View. *The Ecologist.* http://
www.theecologist.org/campaigning/culture_change/375132/plastic_bag_
art_how_to_spoil_any_view.html. Accessed 1 October 2015.

Shanghai Roots and Shoots. 2011. *Plastic Bag Monster for Jane Goodall Institute.*
https://www.youtube.com/watch?v=tsSnVDKn-iU.   Accessed   15
August 2015.

Signs of the Times. 2012. *Vogue Deutsch,* October, 246–255.

CHAPTER 8

Plastics Manufacture and Sustainability: Start Thinking in Cycles

Eric Bischof

Plastics are an unusually versatile materials group, which is found in almost all areas of everyday life, yet products made of plastics are seen by many as resource-consuming, throwaway articles that are not sustainable. Can these very different positions be reconciled? Is it possible to have sustainable plastics at all? This chapter examines this dichotomy from a global manufacturing perspective.

Two aspects must be considered if this objective is to be achieved. Firstly, all three aspects of sustainability: people, planet and profit (Kenton and Berry-Johnson 2020). Secondly, the entire lifecycle of plastics products from their production, processing and consumption to collection and recycling at the end of their service life in the broadest sense.

It is time to start thinking in cycles. New approaches outlined in this chapter can make an important contribution to a more resource-efficient and circular economy, which can include energy recovery as part of the carbon loop.

E. Bischof (✉)
Covestro, Leverkusen, Germany
e-mail: eric.bischof@covestro.com

141

HOW PEOPLE, PLANET AND PROFIT DRIVE SUSTAINABILITY

Manufacturers increasingly focus on aligning their activities with the UN Sustainable Development Goals (SDGs) in support of global progress at the ecological, economic and social levels. Goal 1, for example, aims to end poverty in all its forms everywhere: 'Despite the fact that the global poverty rate has been halved since 2000, intensified efforts are required to boost the incomes, alleviate the suffering and build the resilience of those individuals still living in extreme poverty […]' (United Nations n.d.).

Plastics have an important part to play in this respect. They can help increase food productivity through affordable solutions particularly for farming in developing countries. In Southeast Asia, many people depend on agricultural production, especially those at the bottom of the income pyramid. An estimated 30–50% of their harvested goods, however, are not turned into income (Covestro 2018, 37).

Covestro, a firm leading in the creation of polymer materials, collaborates in this region with customers, NGOs, social enterprises and governmental institutions in food security solutions that use plastics to benefit farmers by reducing post-harvest losses, increasing incomes and improving livelihoods. For example, solar dryer domes, made of high-performance polycarbonate sheets offering special UV-protection, are increasingly used to dry products such as bananas, tomatoes and chilies, significantly reducing the risk of spoilage leading to up to 50% less food wastage. Their parabolic shape allows the best use of solar radiation and water runs off easily. They also address hygiene concerns caused by dust, water contamination and animals and are easy to transport and install (Covestro 2018, 36–38) (Fig. 8.1).

More than 500 of these units have already been installed in Thailand and the dryers are also used in other countries in the region. Their high efficiency is their main advantage. The small amount of energy required for the ventilation system, among other things, is provided exclusively by solar power and the polycarbonate makes for highly efficient insulation. The sun's rays are able to penetrate the unique structure of the material, allowing the dryer to heat up quickly. If the sun shines from dawn until dusk, the temperature inside can climb to 60 degrees Celsius, halving the usual drying time. At the same time, the material filters ultra-violet rays, allowing the product's natural colour to be retained for the most part, a major advantage when selling to the dried snacks market (Covestro 2018, 39). In another model, the cover is made also of polycarbonate but with

Fig. 8.1   Solar dryers help small farmers reduce the risk of spoilage and improves their livelihood. Photo: Covestro

polyurethane applied inside. These dryers reduce moisture in agricultural produce, enabling farmers to preserve their products for up to a year and earn additional income through sales of dehydrated products. The technology is based on conduction as the mode of heat transfer, leading to enhanced efficiency.

Cold chain solutions also based on plastics' insulating potential pursue a similar goal. For example, the primary aim of the Indian start-up, Ecozen, is to enable rural farmers to keep their products fresh without the use of chemicals. It addresses these post-harvest problems by provision of solar-powered micro cold storage facilities to horticulture farmers and other cold-chain stakeholders. They enable the harvest to be stored in optimum conditions thereby increasing its shelf life. Traditionally, such systems have required connection to the electrical grid and have therefore usually only been accessible in urban areas and to large farming cooperatives, leaving out smaller farmers and forcing them to sell their products at low prices.

Experts agree that the SDGs have significant economic potential and could lead to profound changes and above-average growth in many industrial sectors (Covestro 2018, 82). In the plastics industry, this impact could be especially significant. For manufacturers, the economic potential is an important argument for further advancing the issue of sustainability. Simply put, delivering on societal needs will yield economic returns in the long run. Economic solutions on the other hand will attract capital, which in turn leads to scale-up, and this helps a larger number of people. That is why 'people, planet and profit' is the right mindset to create scalable solutions.

Our world is constantly changing and we have increasingly to cope with enormous challenges. With this shifting background, manufacturers are striving to make plastics the solution for many of these challenges. One trend to consider is the combination of population growth and increasing urbanisation. The world will have around 8.6 billion people in 2030, according to the Population Division of the Department of Economic and Social Affairs of the United Nations (United Nations 2017). Alongside this, more than 50% of today's population lives in cities, a shift that has become more pronounced in the last 60 years or so. By 2050, 68% of the world's population is projected to be urban dwellers. Also, by 2030, the world is expected to have 43 megacities, with most of them in developing countries such as India, China and Nigeria. The United Nations says the sustainable development of these megacities depends increasingly on the successful management of urban growth, and that it is important to focus on the need for housing, education, health care, decent work and a safe environment (United Nations 2018).

Manufacturers are already providing solutions based on plastics, which help to improve food supply. Today's domestic refrigerators contain plastics not only in the products that are kept cold inside, such as bottles, bags and packaging, but also in the refrigerators themselves. Plastics are key to their energy efficiency. Thin layers of foamed polyurethane between moulded panels of polypropylene form a protective barrier to help keep their interiors cold. The excellent thermal insulation properties of rigid polyurethane foam make it the material of choice also for insulating appliances such as freezers and water heaters. The foam has low thermal conductivity and so prevents both heat loss and heat gain, contributing to sustainability.

These insulating qualities can also be applied to the walls of industrial and refrigerated buildings, cold stores, warehouses, exhibition and sports halls, schools and other large buildings. The polyurethane foam not only provides thermal insulation but also meets other technical requirements such as stability, inexpensive assembly, freedom from maintenance and easy disassembly and reconstruction. Polyurethane metal-faced sandwich panels are used in prefabricated external wall and lightweight roof components, where a rigid foam core joins the metal panels together in a shear-resistant manner. Polyurethane is also increasingly being used as the core material of load-bearing sandwich panels for refrigerated vehicles. Without question, rigid polyurethane foam makes an important contribution to securing global food supplies, and about 95% of the world's refrigerators are thermally insulated with it. The latest technology in this field is a plastics foam with micro pores that reduces weight and improves insulation even further.

New developments build on existing know-how. Refrigerators being developed by Coolar, a Berlin-based start-up, will use polyurethane as an insulator. However, instead of being powered by electricity, they will use untapped heat from sources such as solar thermal or district heating resulting in refrigerators with a nearly neutral carbon footprint. The heat is used to periodically regenerate an absorbent, which in turn is needed to support cooling by evaporation. Key, however, is a high-performance insulation. With this combination of technologies, cooling to even below freezing point is possible. The technology can be fitted in small cooling systems giving these units particular potential for use in, say, hospitals in developing countries, where there may be no electrical power (and hence no conventional refrigeration). The adequate storage of vaccines and medicines remains a problem in these areas of the world. The development and production of innovative appliances such as those from Coolar will rely on the use of an optimum combination of materials, and the start-up is currently comparing suitable polyurethane systems for insulating purposes (Covestro 2018, 47).

Similarly, polyurethane makes a valuable contribution to reducing energy consumption and finds many applications in the construction sector, the largest sector in terms of energy consumption, accounting for over one-third of final global energy consumption. If no action is taken to improve energy efficiency in buildings, energy demand is expected to rise by 50% by 2050. The International Energy Agency points out that the building envelope (the physical separator between the conditioned and unconditioned environment of a building) determines the amount of energy needed to heat and cool and needs to be optimised to keep heating and cooling loads to a minimum (International Energy Agency 2013, 1–3).

The construction industry is therefore key for plastics manufacturers. The leading players in plastics have a strong presence in the sector and are helping to meet a growing demand for sustainable solutions in buildings based on plastics products and ingredients. The specific properties of some plastics materials allow for unusual solutions. Modern insulation materials can contribute significantly to affordable living spaces in developing economies.

For affordable housing, the walls can be made of rigid polyurethane foam, a material that is extremely stable and has excellent insulating properties, making it an effective substitute for concrete or bricks. Not only does this enable buildings to be erected at low cost and extremely quickly but also ensures they are particularly energy efficient. Additionally, these affordable houses are commercially viable and profitable for all stakeholders along the value chain. The buildings can also withstand high wind speeds and are therefore suitable for natural disaster relief areas.

The future of urban planning is very much open to new ideas based on sustainable living via plastics. The Chinese government is promoting sustainable urban planning concepts and a wider use of renewable energies. In 2018, a student competition called 'Solar Decathlon' was held for building concepts that use only solar energy; 22 student teams from 37 universities from around the world took part in the competition. The venue for this 'Olympic Games for solar energy and green building industry', as the organisers described it, was the city of Dezhou, also known as 'Sun City' because it is at the centre of the Chinese solar industry (Solar Decathlon China n.d.). Each team was required to design and build a one- or two-story solar house with a floor space of between 120 and 200 square metres. The solution was evaluated according to the criteria of architecture, market appeal, engineering, communications, innovation, comfort zone, appliances, home life, commuting and energy performance.

One entry, the *EnergyPLUS Home 4.0*, was from Tongji University of China in collaboration with TU Darmstadt of Germany (Lyu Yan et al. 2017). It had a modular structure so that several units could be combined to form an apartment building, as well as transparent and heat-insulating polycarbonate panels for the façade, and polyurethane raw materials for water-based, yet water-permeable outdoor floor coatings. They are a good fit for new city drainage systems like 'sponge grounds' that soak up water and direct it into underground piping systems.

Another entry from the Indian Mumbai's Shunya team, *Solarise*, combined regional architectural principles and included climatic and cultural aspects. Insulating polycarbonate panels were used on the roof and polyurethane insulation panels in the walls and floors. The concept was a net positive energy solar villa with a floor space of 150 square meters, suitable for a family of six people, comprising all modern amenities for comfortable and sustainable living (Team Shunya n.d.).

The need for advanced building materials is not restricted to developing countries, of course. Neither should they be seen as a low-quality solution for such countries. In industrialised countries like Germany, affordable and energy-efficient accommodation is also in short supply, especially in major cities. At a refugee camp at Bergisch-Gladbach near Cologne, for example, it was shown how alternative construction methods can be applied in developed countries. Here, Covestro, a manufacturer of high-tech polymer materials for key industries, has joined forces with the French construction company, Logelis, and local authorities to provide high-quality accommodation within this refugee camp (Covestro 2018, 39).

Energy-efficient solutions are also highly important in the automotive industry as demographic change and population growth continue to fuel demand for greater mobility. Today, there are already over 900 million passenger vehicles around the world. In highly developed economies, there is approximately one car for every two people; in developing countries, this figure is about one for every ten people (Shell Deutschland/ Prognos n.d.). With increasing mobility in an ever-more urban environment, it is an entirely valid question to ask how the automotive sector will develop in the future and, more specifically, how it can develop sustainably.

New forms of mobility call for innovative solutions. Materials with low weight are particularly in demand in the automotive industry, especially in the promising electric mobility market. Polycarbonates, for example, are convincingly lightweight and versatile, and also offer characteristics such as high-quality optics and transparency. This is why manufacturers are increasing their activity in this area and working on technological innovations such as wrap-around polycarbonate glazing instead of windows or fibre-reinforced composite materials. These materials can replace steel and aluminium in various vehicle components, with weight reductions in the components of up to 50% (German Chemical Industry Association (VCI) / Deloitte n.d., 18). This leads to reductions in fuel consumption and CO_2 emissions or, in the case of electric cars, extension of their range.

Environmentally friendly vehicle repair coatings are a further innovation. An automotive refinish coating with a bio-based hardener, resulting in a reduced environmental footprint, was used in the 'Sonnenwagen', a 200 kg solar-powered vehicle, made of numerous high-tech materials, built by students from Aachen University in Germany. During the 2017 World Solar Challenge, it travelled thousands of kilometres in difficult weather conditions through the Australian outback, without a single drop of fuel. This demonstrated what is possible already today and how high-tech materials can help shape future mobility (Covestro 2018, 32–33).

By contributing to reducing energy consumption in buildings and fuel consumption in vehicles, high-tech plastics provide solutions for major global challenges. They particularly comply with two Sustainable Development Goals: 11—'Make cities and human settlements inclusive, safe, resilient and sustainable' (United Nations n.d.) and 13—'Take urgent action to combat climate change and its impacts' (United Nations n.d.). The examples cited above show that plastics manufacture can help solve many of the fundamental challenges of our time while simultaneously limiting their environmental impact, ultimately allowing us to live within the regeneration limits of our planet. At the same time, it is important that these solutions are economically feasible, as only such solutions can be installed in sufficient quantity, without being dependent on donations or subsidies, and thus reach a scale that makes a real difference to the world.

Perhaps additional encouragement can come from Young Champions of the Earth, a programme organised by UN Environment (UN Environment n.d.). It aims to identify, support and celebrate outstanding individuals aged 18–30 who offer major suggestions to protect or restore the environment. 'It is imperative that we help create more space for innovation, and that also means demonstrating that there is a huge market and enormous potential for entrepreneurs working on sustainable solutions,' says Erik Solheim, executive director of UN Environment (Covestro 2018, 32). He is also convinced that high-tech polymers can contribute to preserving the environment and shaping the future of our planet: 'Plastics and polymers are, and will be, at the very heart of an inclusive, healthy and low carbon economy. The sector is where some of the most exciting green innovations are taking place'. He added that while single-use plastics and the enormous waste that comes as a result are a substantial concern, the problem is not plastics itself, but what is done with them (Covestro 2018, 25).

The advantages of plastics lie in their many advantageous properties, including hardness and strength, lightness, elasticity, temperature resistance and chemical resistance. They are also extremely variable in the form they can take, depending on the manufacturing process, starting material and additives. This is why plastics can offer creative answers for so many of today's challenges.

FINDING THE ACHILLES' HEEL OF PLASTICS

Plastics products differ widely in their properties and their life expectancy. Some can be used for less than one year, others more than 15 years and some even 50 years or more (PlasticsEurope 2017). However, plastics do become waste when their intended use comes to an end, be this after a variety of years in one application, after multiple recycling cycles or after single use. In general, too many plastics products are used only once before they are thrown away. According to estimates, only 5% of the value of plastics packaging material remains in the economy while the rest is lost after a short single use. According to the European Commission, the annual costs range from EUR 70 to 150 billion (European Commission 2018).

One of the most concerning aspects is the amount of plastics waste that ends up in the oceans. Their impact on human health and the environment are among people's biggest environmental concerns: 74% of Europeans are worried that marine litter might be detrimental to health, while 87% are concerned about environmental threats (European Commission 2018).

Single-use plastics are especially likely to become litter in the environment. The most commonly found items on beaches are cigarette butts, drinking bottles, or sweet wrappers. According to recent scientific data, single-use plastics represent half of all marine litter, and particularly concerning are the forecasts: by 2050, there could be by weight more plastics than fish in the oceans (European Commission 2018).

The leakage of plastics, the unintended migration of waste into the environment, is not acceptable and must be stopped. Plastics can take hundreds of years to decompose. If leakage is not prevented, the credibility and acceptance of plastics solutions will be lost and possible solutions to some of the major global challenges will not be available. It is crucial that suitable waste management and consumer education are introduced globally.

The European Commission has committed itself to have all plastics packaging reusable or recyclable in a cost-effective manner by 2030, and a new EU-wide strategy on plastics has fixed a target of 55% recycling of plastics packaging waste by 2030 (European Commission 2018, 11). These efforts are expected to create hundreds of thousands of new jobs in the sorting and recycling industries (European Commission 2018, 2).

A significant portion of plastics waste in Europe is still sent to landfill. In 2016, about 27 million tons of plastics waste was collected through official schemes in the EU 28 member states (plus Norway and Switzerland) in order to be treated. While roughly 31% of this waste was recycled, 27% went to landfill (PlasticsEurope 2017, 30). This situation should be avoided from the manufacturer's point of view: plastics are too valuable to simply be dumped. The plastics industry has made a voluntary commitment in this area, focusing on (1) increasing re-use and recycling, (2) preventing plastics leakage into the environment, and (3) accelerating resource efficiency (PlasticsEurope 2018 Press Release).

Plastics that cannot be recycled must be incinerated in order to at least recover their inherent energy. Today plastics are largely based on fossil fuels and burning plastics causes CO_2 emissions, which in turn is a major contributor to climate change. Yet, as long as we burn fossil fuels to generate energy anyway, we can replace such burning by using plastics in their place; plastics that have had a useful first life. With this approach we can on the one hand reduce the environmental impact of landfills, while on the other help conserve increasingly scarce fossil fuels such as crude oil.

But the problem remains: what happens when we have shifted our energy supply away from fossil fuels towards regenerative energies? Burning plastics based on fossil fuels will then no longer be acceptable.

THE FUTURE BELONGS TO ALTERNATIVE RESOURCES

So it is clearly time to start thinking in cycles. In any discussion about sustainability, the entire lifecycle of plastics products must be taken into account. Benefits in one area must not be offset in others. The aim is to identify weak points and work on solutions, solutions that must be sought in production as well as during and after service life.

In order to address the second of the two Achilles' heels of plastics, we must find reasonable alternatives for their production, aiming to avoid the use of carbon from fossil sources. How can this be achieved? The chemical industry already uses alternative carbon sources such as plants for plastics production. Aniline, for example, a crucial feedstock for plastics, can now also be manufactured from industrial sugars rather than crude oil. Aniline is used to manufacture methylene diphenyl diisocyanate (MDI), a key component in the production of rigid polyurethane foam. This bio-based solution, developed by Covestro, is still on laboratory scale but it holds much promise as an important solution in the future (Covestro 2018, 29).

Another example is polyurethane hardeners containing 70% renewable carbon from cornstarch. The hardeners are used for coatings and adhesives, and a clear coat based on them has already been applied on a vehicle from the Audi Q2 series. There is a similar product for the furniture industry containing 66% bio-based carbon (Covestro AG 2018, 30).

But even CO_2 can be used as a new raw material: in 2016, an initial production plant came on stream where manufacturers succeeded in replacing up to 20% of the traditional fossil fuel based raw materials by CO_2 in one of their product families. The CO_2 is chemically bound and cannot leak. This is a particularly sustainable method of producing flexible polyurethane foam for use in mattresses and upholstered furniture. Research on other CO_2-based applications is ongoing (Covestro 2018, 28).

In parallel, a consortium of 14 partners from 7 countries is planning to investigate how flue gas such as CO_2 and carbon monoxide from the steel industry can be used to produce plastics in a particularly efficient and sustainable way. This will also save crude oil, the raw material used in conventional methods. The cross-sector project called 'Carbon4PUR' receives funding from the European Union and will provide answers on how to significantly reduce the carbon footprint in the future (Covestro 2017).

In addition, post-consumer plastics waste can be used as a source of carbon. Pyrolysis of mixed plastics waste can lead to mixtures of chemicals that can in turn be purified in refineries, ultimately generating basic organic chemicals. These can be used to make new products (Citeo et al. 2019; Plastic Energy 2020).

It is becoming very clear that the future belongs to alternative resources and the urgency for such a transition is bigger than we may be aware. As economies shift their energy production away from fossil fuels, the feedstock availability for the chemical industry will fundamentally change. 'Substituting biomass for coal, electricity and CO_2 for crude oil, biogas for natural gas—the current raw materials debate seems to center on replacing one type of fossil raw material with a renewable resource wherever possible. But this creates a distorted picture of the scale of the challenge ahead: it is extremely unlikely that the value chains in ten, 30 or 50 years' time will be the same as those we know today,' says Prof. Dr. Kurt Wagemann, executive director of the German Society for Chemical Engineering and Biotechnology (DECHEMA) (Covestro 2018, 31).

The need for greater efforts to broaden the raw material base of the chemical industry was also underscored at the Berlin raw material summit which took place, for the second time, in June 2018. The event was co-organised by DECHEMA, the Technical University of Berlin (TU Berlin) and Covestro. 'There are a number of promising approaches when it comes to producing chemicals largely without the use of crude oil,' summarised Prof. Dr. Reinhard Schomäcker from the Institute for Technical Chemistry at the TU Berlin. 'Close collaboration between research-oriented companies and application-oriented science is vital in this regard' (Covestro 2018, 30).

INNOVATIVE APPROACHES TO ACHIEVE A CIRCULAR ECONOMY

While innovation, research and close cooperation with the manufacturing industry is an important factor in improving the production and performance of plastics during service life, it is also vital to utilise all possible options at the end of service life. Different approaches are already being taken to achieve a more circular economy. The chemical industry focuses on different measures: take-back business models improved by the analyses of customer data, an efficient harmonisation of waste capture and logistics, sorting and/or treatment, and subsequent utilisation, and the recovery of energy (German Chemical Industry Association (VCI) / Deloitte n.d., 16–21).

For instance, an innovative return system can be found for solvents in industrial cleaning. The provider SafeChem offers its customers a leasing model for solvents, for manufacturers of cleaning machines, chemical dealers, and waste disposal companies. SafeChem has been able to achieve a reduction in the proportion of solvents in waste water by up to 80%, a reduction in the health risk for employees, and a reduction in the quantity of the replacement solvent also by up to 80% (German Chemical Industry Association (VCI)/Deloitte n.d., 18).

Another example of chemical recycling is the Waste2Chemicals initiative. This is a consortium of eight international companies, including Enerkem, Air Liquide, and AkzoNobel, which plans to start a joint production of bio-based methanol and ethanol from municipal waste. The technology is compatible with existing waste infrastructure and is intended to enable waste that cannot be mechanically recycled to be converted into fuels and high-quality chemicals (German Chemical Industry Association (VCI)/Deloitte n.d., 18).

The different measures and approaches complement each other. There cannot be a single solution that will solve everything at once; each route has advantages and disadvantages and so the ultimate solution will be a combination of methods, ultimately compatible and possibly synergistic. Maintaining, reusing, refurbishing, recycling, and energy recovery are ultimately concentric circles designed to fulfil the intended purpose with minimal resource consumption and waste disposal. At the centre of these concentric circles is a long-lasting product. For products with only shorter time spans, recycling can extend the lifespan of the underlying plastics material by reusing it many times, although at some point its properties will deteriorate and its usefulness will end. In some cases, mechanical recycling may not be feasible. Take, for example, composites, these materials can provide unique combinations of properties but may make mechanical recycling more difficult. Pure one-component systems are better for recycling. There have been calls for legislation in this direction, but this would not take all options into account.

Let us suppose there is a need for a material that must fulfil two requirements, for example, a transparent wrapper that provides an oxygen barrier and is stable enough to cover food for supermarket display. Two different materials might be considered. A one micrometre thin film would be a good oxygen barrier; however, it would have to be 100 micrometers thick to achieve the required mechanical stability. The other material is mechanically stable at one micrometre thickness but requires 100 micrometres to be a sufficient oxygen barrier. Combining the two would allow creation of a two micrometres thin composite, effectively reducing the resource consumption by a factor of 50. In comparison, you would have to recycle the alternative one-component system 50 times just to break even and without using any energy to do so. This, of course, would not hold true in every case, but composite solutions should not be undervalued. Rather the performance improvement and energy efficiency they are able to provide in specific applications should be embraced.

Generally, energy stored in plastics can be recovered. Burning is an elementary component of sustainable future concepts. Thermal energy recovery from plastics waste offsets a corresponding amount of fossil fuel that would otherwise be needed for energy generation. Hence, burning plastics waste for gaining energy is better than burning fossil fuel, particularly if the plastics are not made from fossil raw materials in the first place.

As the energy sector is more and more on its way to regenerative sources, the consequent task of plastics manufacturers is to reduce their dependency on crude oil as a raw material. Burning plastics for energy recovery needs to be combined with alternative sources of carbon (bio-based, CO_2 or a combination thereof). However, two important criteria have to be met: the production of the bio-based raw material needs to be energy efficient, and the performance of the respective product should not be diminished. If these criteria are met, this approach is a contribution to a more circular economy which brings energy recovery back into the carbon loop. After all, CO_2 is a key element of natural processes and of life itself. Where CO_2 can be re-introduced into circular solutions, we are mimicking nature itself.

CONCLUSION

In conclusion, the three dimensions of sustainability, people, planet and profit, must be taken into account in the journey of finding solutions for the future. While important steps have already been made in terms of plastics in use, there is still a demand for focusing on production and end-of-service life. By intelligently combining different approaches to solutions, the advantages of plastics can be utilised to a greater extent while disadvantages can be overcome. Some solutions are already on the market or being introduced, while others will only be available in some years or decades but one thing is certain: to be successful in the end, any solution needs to help preserve the planet, improve the quality of life and safety of millions of people, and be economically worthwhile.

BIBLIOGRAPHY

Citeo, Total, Recycling Technologies, Mars and Nestlé. 2019. We've Joined Forces to Develop Chemical Recycling of Plastics in France. December 10. https://recyclingtechnologies.co.uk/2019/12/citeo-total-recycling-technologies-mars-and-nestle-join-forces-to-develop-chemical-recycling-of-plastics-in-france/. Accessed 20 May 2020.

Covestro AG. 2017. Europeans Join Forces on CO_2. Press release. http://presse.covestro.de/news.nsf/id/Europeans-join-forces-on-CO2. Accessed 18 June 2018.

———. 2018. Annual Report 2017. https://www.google.com/search?client=firefox-b-d&q=Covestro+Annual+report+2017. Accessed 20 May 2020.

European Commission. 2018. *Questions & Answers: A European Strategy for Plastics*. Fact Sheet. Strasbourg: European Commission.

German Chemical Industry Association (VCI) / Deloitte. n.d. Chemistry 4.0: Growth Through Innovation in a Transforming World. https://www2.deloitte.com/content/dam/Deloitte/global/Documents/consumer-industrial-products/gx-chemistry%204.0-full-report.pdf. Accessed 18 June 2018.

International Energy Agency. 2013. Transition to Sustainable Buildings Strategies and Opportunities to 2050. https://www.iea.org/Textbase/npsum/building2013SUM.pdf. Accessed 12 June 2018.

Kenton, Will, and Janet Berry-Johnson. 2020. What Is the Triple Bottom Line. https://www.investopedia.com/terms/t/triple-bottom-line.as. Accessed 12 May 2020.

Lyu, Yan, Yiqun Pan, and Cuisong Qu. 2017. Energy System Design and Optimization of a Solar Decathlon House. *Procedure Engineering* 205: 1019–1026. https://isiarticles.com/bundles/Article/pre/pdf/150957.pdf? Accessed 20 May 2020.

Plastic Energy. 2020. INEOS and Plastic Energy to Collaborate on New Advanced Plastic Recycling Facility. Press release April 21. https://plasticenergy.com/ineos-and-plastic-energy-to-collaborate-on-new-advanced-plastic-recycling-facility/. Accessed 20 May 2020.

PlasticsEurope Association of Plastics Manufacturers. 2017. Plastics – The Facts 2017: An Analysis of European Plastics Production, Demand and Waste Data. https://www.plasticseurope.org/application/files/5715/1717/4180/Plastics_the_facts_2017_FINAL_for_website_one_page.pdf. Accessed 13 June 2018.

———. 2018. Press Release. Plastics 2030. Voluntary Commitment. https://www.plasticseurope.org/en/newsroom/press-releases/archive-press-releases-2018/plastics-2030-voluntary-commitment.

Shell Deutschland / Prognos AG. n.d. Shell Pkw-Szenarien bis 2040: Fakten, Trends und Perspektiven für Auto-Mobilität. https://www.prognos.com/uploads/tx_atwpubdb/140900_Prognos_Shell_Studie_Pkw-Szenarien2040.pdf. Accessed 12 June 2018.

Solar Decathlon. n.d. Solar Decathlon China 2018. https://www.solardecathlon.gov/international-china.html? Accessed 20 May 2020.

Team Shunya. n.d. Team Shunya IIT Bombay. https://teamshunya.in/. Accessed 29 August 2018.

UN Environment. n.d. Champions of the Earth. https://web.unep.org/young-champions/. Accessed 29 August 2018.

United Nations. n.d. Sustainable Development Goals. Sustainable Development Platform. https://sustainabledevelopment.un.org/?menu=1300. Accessed 10 June 2018.

———. 2017. World Population Prospects: The 2017 Revision. Press Briefing. Economic & Social Affairs. http://www.un.org/en/development/desa/population/events/pdf/other/21/WPP2017_press.briefing_Directors-web.pdf. Accessed 12 June 2018.

———. 2018. World Urbanization Prospects: The 2018 Revision. Economic & Social Affairs. https://esa.un.org/unpd/wup/Publications/Files/WUP2018-KeyFacts.pdf. Accessed 12 June 2018.

PART II

Plastics in Societal Use

CHAPTER 9

'Plastic Fantastic Lover': Plastics and Popular Culture … 1945 to the Present

Mark Suggitt

The chapter's title is from that of a song by the 1960s San Francisco acid-rock band Jefferson Airplane. It was written by one of their singers Marty Balin, and appeared in their 1967 album, *Surrealistic Pillow*. I chose it because the meaning of the song, like those many meanings ascribed to plastics, is ambiguous. Many thought it a social commentary on modern life, love and consumerism, others thought it was about a dildo (plastics obviously), while Balin maintained it was a tribute to the new stereo system he had bought with his royalties from shipping thousands of vinyl records for the Radio Corporation of America's RCA Victor label (Song facts n.d.).

Cultural studies tend to consider plastics in relation to other contexts and arguments. This chapter examines rather how plastics have been popularly perceived and the part they have played in shaping and enabling popular culture. In doing so, I have deliberately focused my observations

Mark Suggitt was deceased at the time of publication.

M. Suggitt (Deceased) (✉)
Derwent Valley Mills World Heritage Site, Derbyshire, UK
e-mail: author@noreply.com

© The Author(s), under exclusive license to Springer Nature Switzerland AG 2020
S. Lambert (ed.), *Provocative Plastics*,
https://doi.org/10.1007/978-3-030-55882-6_9

on Britain and the United States, partly due to the limitations of a single chapter, but mainly because these two countries produced the most influential manifestations of popular culture in the last half of the twentieth century. The reasons for their development were both different and complementary; and they both fed off and resisted each other.

The word 'plastic' was certainly a signifier within 1960s counterculture, and it was usually a negative one ... an association with the straight consumerist world. How did it get there? It is not what Victor Yarsley and Edward Couzens envisaged in 1941 in their Pelican book on Plastics:

> When the dust and smoke of the present conflict have blown away and rebuilding has well begun, science will return with new powers and resources to its proper creative task. Then we shall see growing up a new, brighter, cleaner and more beautiful world ... the Plastics Age. (Yarsley and Couzens 1941, 152)

Both Yarsley, a chemist, and Couzens, a research manager at British Xylonite Plastics, were advocates of the material, and like others writing in the midst of a global war, they looked forward to a better future. They concluded their book by imagining the world of the 'Plastic Man', a dweller in the Plastics Age that was already dawning (Yarsley and Couzens 1941, 151–152). It charted his interactions with the material world, from the cradle to the grave and most of the materials he encountered were plastics. Even when we take into account their zeal, it is a remarkably accurate account of how we live today. Not all their predictions, such as plastics aeroplane wings and fuselages, came true although plastics are now increasingly fundamental in aeroplane construction (British Plastics Federation n.d.). However, most did, if not, perhaps, in as universal a way as they envisaged. This was to a large extent a result of post-war cultural conditions.

World War II produced great advances in plastics production and the post-war period allowed commercial companies to capitalise on these. Plastics industries were a part of that post-war reconstruction, especially in Italy and France. The economy of the United States had not been destroyed and benefitted from the war; it was about to enter a period of unprecedented growth and disposable income: the Affluent Society of J.K. Galbraith (1958) (Picketty 2014, 96–99). The United States and Britain were both to innovate in the second half of the century. How did plastics fare in these countries and how were they viewed during this period?

AMERICA

The United States took the lead in the immediate post-war period. America's gross national product rose from around $100 billion in 1940 to over $500 billion in 1960 (Heimann 2014, 19). US plastics production quadrupled during the war thanks to its many wartime uses, from aeroplane and munitions parts to GI issue combs. It rose from 213 million pounds in 1939 to 818 million pounds in 1945, so it is hardly unsurprising that even during the war American companies had a weather eye on the benefits of peace and the uses to which the newer polymers could be put. As early as 1943, DuPont was prototyping housewares made of the plastics that were currently consumed by wartime demands (Freinkel 2011, 25). The first National Plastics Exposition opened at the Grand Central Palace in New York in April 1946. Organised by the Society of the Plastics Industry, which was formed in 1937, it attracted so many people that the fire marshals had to control entry. There were 164 exhibitors displaying their wares in 10-foot by 10-foot booths over 24,600 square feet. All of them were US companies, all ready to sell a plastics version of the American Dream. To Yarsley and Couzens, sitting in bombed out Britain, it must have seemed like a dream. Susan Freinkel described it as:

> … an exciting and glittering preview of the promise of polymers. There were window screens in every color of the rainbow that would never need to be painted. Suitcases light enough to lift with a finger, but strong enough to carry a load of bricks. Clothing that could be wiped clean with a damp cloth. Fishing line as strong as steel. Clear packaging materials that would allow a shopper to see if the food inside was fresh. Flowers that looked like they'd been carved from glass. 'Nothing can stop plastics,' the chairman of the exhibition crowed. (Freinkel 2011, 25)

The Plastics Industry harnessed the power of magazines and the advertising agencies of Madison Avenue to convince American households to buy their products. The magazine *House Beautiful*, an established home interiors magazine, collaborated with the Society of the Plastics Industry to produce a special 55-page issue in October 1947 entitled 'Plastics…a Way to a Better, More Carefree Life'. Such an intervention was timely as the quality of many post-war plastics products was poor and the key decision makers in this area of household spending were women. Plastics were sold, along with other increasingly sophisticated domestic appliances, as labour saving and hygienic. Domestic refrigerator sales rose by 80% during the 1940s, a fact that was bound to aid the sale of plastics containers like Tupperware that featured in the second National Plastics Exposition of 1947 (Clarke 1999, 55). The magazine's editors visited the show and, as Alison J. Clarke noted, the resulting edition served as a consumer hand-book describing the relation of plastics to the intricacies of modern home-making. The potentially alienating aspects of the new materials, unfamiliar in texture, density and form, were countered by revelations regarding their labour saving, wipe-clean features. Tupperware, 'Fine Art for 39 Cents' conjured up a far more elaborate set of cultural meanings (Clarke 1999, 42–44).

This was nine years before Tupperware was selected for display as a modernist icon at the Museum of Modern Art in New York and four years before the birth of the Tupperware Party, which was later to become a pop-cultural phenomenon in itself. Interestingly, the Tupper Corporation's advertising also concentrated on the look and usefulness of the product, especially the seal. It never drew great attention to their chemical origins although a number of adverts did describe Tupperware as 'plastic' products.

Another cultural artefact that reached millions of Americans was the Monsanto 'House (or Home) of the Future', which opened in Tomorrow Land at Disneyland California on 7 June 1957. The house was a collabo-ration between Disney Imagineering, the Massachusetts Institute of Technology (MIT) and the chemical giant, Monsanto, who had already funded the theme park's Hall of Chemistry. Walt Disney was happy to have a popular attraction that fitted into Tomorrow Land, although ironi-cally Disney Studios had produced *The Plastics Inventor* back in 1944. This animated cartoon featured Donald Duck as a modern alchemist attempting to transform household waste into a plastics aeroplane. True to the perception of being cheap and shoddy (and also the fact that the hap-less Donald always courted disaster), his aeroplane disintegrated when exposed to the elements (Clarke 1999, 42).

Funded by Monsanto, it was a scientific investigation into how plastics could be used to produce inexpensive mass housing. It was also a sales pitch. Its designers, Marvin Goody and Richard Hamilton were MIT fac-ulty members sponsored by Monsanto to develop new markets for its plas-tics products; a wise move as by then 15% of US plastics production went into the domestic market. In addition, as far as possible, everything in it, from the furniture, floors, walls, soft furnishings and carpets, was to be made of plastics using the full range of acetates, rayons, nylon and vinyls and so on. It took six months to build at a cost of $1 million, but most of this was spent on research and it was estimated that if produced in bulk it could sell for between $10,000 and $15,000. The futuristic cross-shaped design with fibreglass walls was not practical but it was certainly dramatic. A free attraction, it remained open until 1967 and was eventually demolished. As well as demonstrating the versatility of plastics it also featured a microwave oven, dishwasher, security system and a (dummy) flat-screen television (Weinstein n.d.). The House of the Future chimed with an essentially optimistic, suburban, family-orientated world, which, as exemplified by *The Jetsons*, a futuristic Hanna-Barbera cartoon of 1962, would survive into the future.

As America entered the 1960s, the market for plastics was growing, with, as we have seen, a mixed reputation, they were seen as utopian, cheap and shoddy or like the Tupperware Party, a symbol of suburban American petty bourgeois capitalism. There were the beginnings of dissatisfaction with the post-war settlement, pioneered by the Beat writers and poets of the 1950s. The oil, chemical and plastics industries were a part of Alan Ginsburg's 'Moloch' in his poem *Howl* first published in 1956 (Morgan 2010, 91–147). Not all was well in the fracturing 'American Dream', as exemplified in *Revolutionary Road*, Richard Yates' novel of 1961. Jeffrey Meikle remarked that plastics were a substitute, nothing more, nothing less, and fair game for any critic who considered American life superficial, phony, or abstracted from reality (Meikle 1990, 52). As America entered the age of Vietnam, Civil Rights and a radical counterculture, those critics were to grow in number and express themselves through popular culture. The signs were there in the song 'Plastic Jesus' written by two irreverent teenagers, Ed Rush and George Cromarty in 1957. The essence is that whatever confronts these youngsters, the 'plastic Jesus' on their car dashboard will enable them to go far. The song became a popular folk standard after being published in *Sing Out* magazine in 1964 and reached a worldwide audience when Paul Newman delivered a melancholic version in the 1967 film *Cool Hand Luke* after hearing of his mother's death.

BRITAIN

Meanwhile, life in post-war Britain remained bleak. The country was saddled with huge debts and an exhausted people were struggling to rebuild equally exhausted industries. Unlike in the United States, consumer goods were in short supply. The Council of Industrial Design (CoID) produced the Britain Can Make It exhibition at the Victoria and Albert Museum in 1946. The museum building had survived the war intact but was empty as its collections were still safely stored elsewhere. The CoID was a quasi-autonomous non-governmental organisation established in 1944 with the aim of demonstrating the value of design in reviving post-war Britain. The aim of the show was to feature the best of the best that modern British industry could produce as a 'gesture to the British People and the World' as stated in the Council of Industrial Design's Policy Committee Minutes (CoID 1946 cited in Darling n.d.). The exhibition title referenced the wartime phrase 'Britain Can Take It' but, as all the goods on display where for export and not available at home, the response to the exhibition, although generally popular, was 'Britain Can't Have It' (Bilbey 2019, 27).

The public response to the Britain Can Make It exhibition was recorded in an independent and unique survey by the Mass Observation organisation. It revealed that the public was both discriminating and, as rationing continued, desperate for more consumer goods. However, despite having sections in which plastics could shine, including one on 'New Materials' (Council of Industrial Design 1946) 'antipathy to plastic goods was common, for they still bore the stigma of "cheapness" and "commonness". China, (then still in short supply) was a much more sought-after material' (Bullivant 1986, 150). *Good Housekeeping* magazine noted in its 1957 feature on 'Home Plastics' that:

> After the war, the market was flooded with somewhat indifferent plastic goods, which left many people with a poor impression of the capabilities of this material. In many cases the wrong type of plastic was used for the wrong articles and often they were badly manufactured. (*Good House Keeping*, July 1957, 65–70)

However, there were not that many plastics products on show, although isolated items, including the striking green version of the Wells Coates designed EKCO AD 65 radio were displayed, plastics were mainly represented by rather ordinary products: cheap jewellery, dolls, veneers and so on. Additionally, they hardly featured at all in the recreated rooms, although R.D. Russell's 'living room for a large town house' did have Perspex rods in a sideboard's doors and the section where designers looked to the future did feature one prophetic plastics product: a suitcase with rounded re-enforced edges with a folding carrier. Additionally, a working plastics press featured in a section explaining Industrial Design. It pro-duced 3000 plastics eggcups a day and was a popular exhibit, mainly because it moved (Council of Industrial Design 1946).

By the time of the larger Festival of Britain in 1951, things had not got much better. As Reyner Banham remarked, the Festival was 'Britain catch-ing up rather than innovating' (Banham 1951 cited in Banham and Hillier 1976, 190–198).

Writing in his semi-autobiographical novel published in 1957, *The Ordeal of Gilbert Pinfold*, Evelyn Waugh wrote:

> His strongest tastes were negative. He abhorred plastics, Picasso, sunbath-ing and jazz—everything in fact that happened in his own lifetime. (Waugh 1957, 14)

In contrast, a very different type of Englishman, Richard Hoggart, wrote of the changes he saw in the working-class home of the 1950s:

> Chain-store modernismus, all bad veneer and sprayed on varnish stain, is replacing the old mahogany; multi coloured plastic and chrome biscuit bar-rels and bird cages have come in. (Hoggart 1957, 35)

Yet another contrast came from France. Also writing in 1957, Roland Barthes declared that:

> Plastic has climbed down, it is a household material. It is the first magical substance which consents to be prosaic. (Barthes 2000, 97)

For Waugh and Hoggart, their distaste of plastics crossed the class divide. Waugh was lamenting, as he had done previously in his novels *Brideshead Revisited* (1945) and the *Sword of Honour* trilogy (1952–1961), an elite, aristocratic way of life that he saw falling away. Hoggart felt that the positive traditions of working-class life were becoming debased. Like J.B. Priestley before them, they lamented different worlds that were pass-ing, both in the process of becoming americanised. Barthes on the other hand presented a more optimistic view. Plastics had yet to reach their potential, either as 'buckets or jewels' (Barthes 2000, 97).

The British remarks illustrate understandable reactions against some of the shoddy goods on sale at the time, but they reveal more than that. There was a cultural assumption against modernity. Pure plastics products found it hard to develop in this world. Dubbed cheap and nasty, they were seen as inferior: materials used to imitate the 'real' materials of the world of Hoggart and Waugh. As Claire Catterall remarked, British plastics from the 1950s failed to exhibit the same verve and grasp of modernity that can be found in parallel products emerging from European countries such as Italy, Germany and Sweden. The reasons for this were more cultural than technological (Claire Catterall 1990, 73). Nevertheless, the uses for plastics kept on growing, especially in the kitchen. A former British Home Secretary, Alan Johnson, recalled this in his impoverished Notting Hill home of the 1950s. When his sister's friends came round:

> they would have to sit in the kitchen, with its four battered chairs round a Formica table (we were big on Formica), one old armchair and the washing line slung across the room under the flyblown lampshade. (Johnson 2014, 97)

Towards a Plastics Age

If things were slow in the world of popular British consumption, life in the fast lane of high design was a lot better for plastics, and that road was more travelled in Europe and the United States, as can be seen by the products designed by Joe Colombo, Verner Panton and Dieter Rams. By the mid-1960s, Britain was catching up, largely due to the work of designers like Robin Day and Kenneth Grange and emerging young fashion designers like Mary Quant who took advantage of the rising quality and versatility of materials. Plastics were proving their worth in the automobile and aerospace industries. In the home they were winning in the proscribed area of the kitchen and the food that entered it was increasingly wrapped in plastics. More and more household and personal goods, from heaters and hair-dryers to cameras and radios were enveloped in plastics cases.

Andy Warhol celebrated the fluidity and ambiguity of plastics in his *Exploding Plastic Inevitable*, the drug-fuelled multimedia events of 1966–1967 that featured music by the Velvet Underground (Torgoff 2001, 156–165; Violet 1988, 101–107; Savage 2015, 189–232). He stated,

> A Pop person is like a vacuum that eats up everything, he's made up from what he's seen. Television has done it. You don't have to read anymore. Books will go out, television will stay. Movies will go out, television will stay. And that's why people are really becoming plastic, they are just fed things

and are formed and the people who can give things back are considered very talented. (Warhol cited by Wasserman 1966)

However, the prevailing mood from a range of cultural commentators was negative. Plastics were often guilty by association. If Warhol and the Jefferson Airplane song were ambiguous ('No-one's wise to my plastic fantastic lover'), others were not. Frank Zappa and the Mothers of Invention's second album, *Absolutely Free*, also came out in 1967. Its first track was 'Plastic People', an incisive commentary on all things American. A collage of spoken word, discordant guitar lines and the riff from *Louie*, it clearly saw plastics as a metaphor for conservative America.

A fine little girl
She waits for me
She's as plastic
As she can be
She paints her face
With plastic goo
And wrecks her hair
With some shampoo

Plastic people
Oh baby now ...
You're such a drag

The sentiment was echoed in 1968 in the Steve Miller Band's song 'Living in the USA: We're living in a plastic land'. Two years earlier, on the other side of the Atlantic The Who's song 'Substitute' contained the line: 'I was born with a plastic spoon in my mouth ...' the polar opposite of the phrase born with a silver spoon in your mouth. The song's protagonist is a misfit, raging against superficiality, duplicity and social class. 'Substitute your lies for fact, I can see right through your plastic mac'. In March 1969 that great observer of quietly desperate Englishness, Ray Davies of The Kinks, concluded that Plastic Man had:

... a phoney smile that makes you think he understands,
but no-one ever gets the truth from plastic man.

Like Davies, The Who's Pete Townsend used plastics as a metaphor for inauthenticity, a substitute material. The myth of cheap and nasty would have been easily understood by an audience growing up in the 1960s, so it is no wonder that 'Substitute' was covered by the Ramones, the Sex Pistols and Blur. 'Substitute' was a big hit but The Kinks' 'Plastic Man' was less fortunate, only reaching Number 31 in the UK Charts. Not one of their better songs, it was not helped by the BBC refusing to play it as it included the offending word 'bum': 'Plastic legs that reach up to his plastic bum'. Both The Who and The Kinks had emerged from traditional English working-class culture through the launch pad of art school and yet they perpetuated the cheap and nasty image of plastics as inauthentic substitutes. They were the heirs to Richard Hoggart, although ironically enough all the equipment both bands used, from amplifiers and drum skins through to guitar picks, relied on increasingly high-performance plastics. Plastics played a key role in the delivery of their music.

Plastics featured also as a metaphor for the straight world in the 1967 film *The Graduate* starring Dustin Hoffman. Based on Charles Webb's novella of 1963, the following lines are by the screenwriter Buck Henry, a man with a dry sense of humour who went on to work with Mel Brooks and Steve Martin. In a scene that has become famous for defining the 'generation gap', Mr McGuire, a family friend, is trying to get the aimless Benjamin Braddock to think about his future.

> I want to say just one word to you. Just one word.
> Yes sir.
> Are you listening?
> Yes I am.
> Plastics.
> Exactly how do you mean?
> There's a great future in plastics. Think about it. Will you think about it?

Clearly, he did not. Again, plastics were seen as part of that tarnished, inauthentic American Dream, which its youth, politicised by the Vietnam War, rebelled against.

Benjamin had no interest in corporate America but he had probably read comics; maybe even *Plastic Man* (Wright 2000, 66–67). Plastic Man was a popular super-hero character with a body that could stretch every which way. He first appeared in the anthology *Police Comics* published by Quality Comics in 1941. Created and drawn by Jack Cole, a highly original comic book artist, Plastic Man began as a small-time crook that gained his powers by falling into a vat of chemicals whilst robbing chemical works. He also changed his ways and used his powers for good. Plastics were part of the funnies and not to be thought about too seriously. A similarly stretchy hero was to re-emerge in the form of Mr Fantastic. He was a member of *The Fantastic Four*, created by Marvel Comics writer Stan Lee and artist Jack Kirby in 1961, which sold so well it revived Marvel's fortunes and convinced Lee to stay in the comics business (Howe 2012). So, along with the other members of *The Fantastic Four*, Invisible Woman, the Thing and the Human Torch, a new version of Plastic Man helped sustain the creators of a world of super-heroes that now realistically strode the world in colossally successful movies thanks to computer-generated imagery.

Ironically, none of these forms of social commentary could have existed without plastics. The technology that allowed the 45 rpm 7″ single to develop was pivotal to the growth of rock n' roll. It was lighter and more portable than shellac 78s (shellac itself a natural plastic) but its sonic limitations meant that the new songs had to be 3-minute masterpieces. Smart producers like Sam Phillips and Phil Spector knew their work had to sound good on a 7″ single played on a portable turntable or a car radio. The LP (long player at 33 1/3 rpm) was introduced in 1948 and paved the way for more thematic albums like those pioneered by Frank Sinatra on Capital Records in the 1950s. Records became 'albums' of songs and some, such as the mid-1960s output of The Beatles and Bob Dylan, became cultural landmarks. From the early 1970s onwards, global album sales became huge.

Likewise, in visual culture, mass-produced copies of movies on celluloid allowed them to also become global phenomena. By 1969, 'plastic' was the word used to name John Lennon and Yoko Ono's musical project: the Plastic Ono Band. Ono stated that the name was coined by Lennon when she told him about a proposed performance where she intended to use four tape recorders on plastics stands as her band (Ono 2010). Lennon and Ono would have been aware of Warhol's project and probably delighted in the ambiguity of the name as the band was intentionally a conceptual movement and had a constantly changing membership from its inception. It had genuine plasticity.

Architects and designers continued to experiment with the boundaries of the material. In Finland, Matti Suuronen designed his 'Futuro' house of 1968 (Gossel and Leuthauser 2005, 376–379). Originally conceived as a portable ski lodge for a friend it achieved its flying saucer shape through the use of glass fibre reinforced polyester (GRP). Suuronen expanded the design into a portable house intended for mass production. In Britain, Lancashire County Council's architect's department designed a portable classroom, also capable of mass production. Designed by Ben Stevenson and Mike Bracewell using an early version of computer-aided design, it was the first fully structural plastics building in the United Kingdom formed of 35 white GRP panels to form an icosahedron. Known as 'The Bubble', the prototype was installed at Kennington Primary School in Preston and remains in use today (Brook 2017). 'The Bubble', 'The Futuro' and Suuronen's later, more conventional 'Venturo' house of 1971 had the potential to offer exciting and workable alternatives in the spirit of the 'House of the Future' but met with a conservative market and fell victim to the 1973 oil crisis that made production costs uneconomic.

The late 1960s and early 1970s began to witness the decline of the brave new world. After the end of the hippie dream, the oil crisis and nuclear tensions, elements of popular culture began to look backwards. Britain saw a revival of interest in the Victorian period and the art deco of the 1930s. American movies such as *Bonnie and Clyde* (1967), *The Godfather* (1972), *The Sting* (1973) and *China Town* (1974) were set in the recent past. Shops like Biba cheerfully plundered from these decades. These factors led to an increasing interest in earlier, stylish plastics products. However, plastics were also becoming part of everyday life. Cars now had plastics seats (which became uncomfortably hot in the sun), teenagers bought records and covered their school-books in vinyl and children fired the tops of Smarties tubes at each other.

Plastics and Punk

The years 1976 and 1977 saw the next seismic shift in popular culture with Punk. Although initially a musical expression, it was much more than that. Punk was both a fun rebellion and also an acceptance of being beaten down by failed systems and failed economies. It also used plastics as a metaphor, but appropriated it to be a symbol of that decline and therefore celebrated. One of its true stylists took the name Poly Styrene. The symbols like the razor blade, the dog collar, the toilet chain, PVC, Lurex and the plastics bin-liner, all worn by early punks were used to portray a new set of nihilistic values, which said, 'we know life's rubbish but we get round it'. Punk used these despised materials as 'confrontation dressing'. Like the Dadaists they spat on everything, including themselves (Hebdige

1979, 107–108). Twenty years later, Kevin Moore, in his book *Museums and Popular Culture*, used the black plastic bin liner as an example of how museums could (if they chose to display them) give different values and uses to such objects. It could:

> be immediately disposed of, with its contents, falling into the rubbish. Alternatively, in 1977 it could have been customised as a piece of clothing by a punk, becoming a spurious masterpiece. This could then be thrown away, again falling into the rubbish. If it was kept as a souvenir, in the 1990s it could have been displayed at the 'Streetstyle' exhibition at the Victoria and Albert Museum as an authentic masterpiece, or in a temporary social history display on youth culture as an authentic artefact. (Moore 2000, 74)

Like all youth movements, these would become identikit uniforms for the followers whilst the innovators had moved on (Cohn 1989, 268–269). For example, the band The Slits rejected a pink plastics sleeve for their 1979 album *Cut*, and went for a much earthier option, covering themselves in mud (Albertine 2014, 210; Westwood and Kelly 2014, 133–213). But Punk's acceptance of the material as a part of a creative bricolage had an effect. Plastics were cooler and Punk's mix-and-match aesthetic went on to influence designers and consumers in the 1980s. The UK group The Buggles continued the metaphor of plastics as symbols of artificiality in their album *The Age of Plastic* in 1980. The song *Living in the Plastic Age* mixed rat-race anxiety with plastic surgery.

> Living in the Plastic Age
> Looking only half my age
> Hello Doctor lift my face
> I wish my skin could stand the pace.

Plastics and Cultural Signifiers

Museums, which are usually behind the curve in respect of shaping taste, had begun to respond. A pioneer in this field was MoMA in New York with its design collection, established in 1932 under the directorship of Alfred Barr, acquiring Tupperware as early as 1947 (email to the editor from MoMA Architecture & Design Study Center, 6 May 2020). The V&A only began acquiring plastics in the 1960s and then largely through the gift of winning designs of the CoID 'Design of the Year' awards discussed in Chap. 12, thus by default rather than from an interest in plastics materials as is further discussed in Chap. 13. In the late 1970s, the Science Museum benefitted from two large donations of plastics from private collectors (Mossman 1997, 12) but interestingly again, thus, not as a result of a proactive interest in plastics by its curators. Plastics had become so much part of day-to-day culture that inevitably they began finding their way into museums. Other British museums began to explore their history: from Wakefield, Salford and Wolverhampton in the 1970s and 1980s through to the Science Museum's *Plastics Gallery* in 1986 and the major *Plastics Age* show at the V&A in 1990. Museums of Social History began to discover that they had plastics objects in their collections, often collected not for their material, but for their function, such as a camera or a cigarette case.

Plastics also began gaining a literature thanks to the work of writers like Sylvia Katz (1984) and Susan Mossman (1997). The British Bakelite Society formed in the early 1980s, followed by the Plastics Historical Society in 1986, which brought together the industry, collectors and museums. *The People's Show* at Walsall Museum and Art Gallery in 1990, exhibited a variety of private collections including one of plastics frogs, egg cups and plastic bags (Windsor 1994, 51).

In the post-punk world of 1986, Cornerhouse in Manchester produced a show entitled *Our Domestic Landscape*, which looked at craft and design in the home. Writing in the accompanying book, Daniel Weil, renowned for his 'Radio in a Bag' put into manufacture two years earlier, argued that:

> The next stage will be to use the material in its own right, and will require a re-assessment of the design of most familiar objects whose logic derives from outdated technology and craft or materials of the Industrial Revolution. (Weil 1986, 31)

The 1980s and1990s saw a new wave of designers, many totally at ease with what they saw as diverse materials that could be used for jewellery, technology, and furniture, whatever. A designer could make a plastics bracelet which looked as good as a gold one, in fact you could combine the materials and produce something new and beautiful. It could now be a jewel and a bucket or an iPhone. Writing in 1984, Sylvia Katz stated that:

Far from imitating traditional materials the designers of 'Objects for the Electronic Age' use the potential of sheet plastic and metal to challenge the way we expect things to look. (Katz 1984, 125)

She illustrated the point with examples of the work of Memphis, Daniel Weil, Kartell and Cassina. Such designs featured in the growing number of 'style guides' like *Elle Decoration*, *The World of Interiors*, *Wallpaper* and the Sunday supplements that aided a greater acceptance of 'modern' furnishings. The arrival of the Swedish furnishing giant IKEA in the UK in 1987 (and the shrinking size of new-build British flats and houses) accelerated this trend and also made it more affordable. It opened its first store in Warrington, Lancashire, which soon became a place of pilgrimage for people from all over the UK looking for affordable, stylish furniture. This success led to the opening of a further 19 stores. The 2018 catalogue features a large array of plastics products including a polypropylene armchair and recycled plastics dining chairs which both retailed for under £70.00 (IKEA 2018, 164, 241).

The fact that IKEA is selling recycled plastics is testimony to the rise of the green movement and concerns over the environment. Awareness of the problems that non-degradable plastics produce has seen another shift in popular culture. We knew they were useful and we accepted them, in some cases as a necessary evil, like the fossil fuels that produced them. They shaped and moved the modern, globally connected world around. Popular music responded, plastics were no longer a metaphor, but directly under attack for being over-produced, non-degradable and irresponsibly disposed of. The singer Damon Albarn's project Gorrilaz produced the album *Plastic Beach* in 2011. The song *Plastic Beach* included the lines:

It's a Casio on a plastic beach
It's styrofoam deep sea landfill.

In 2015, the singer and producer Pharrell Williams collaborated with the fashion company G-Star to produce clothes made from recycled plastics, yarn sourced from waste recovered from the sea. How much impact the well-meaning intentions of artists have had is hard to quantify, but their actions are in tune with those of governments who have introduced compulsory charges on the sale of plastic shopping bags.

CONCLUSION

In conclusion, the development of post-war plastics, defined by misunderstanding, derision, limited acceptance, deviance and qualified acceptance has been both mirrored and shaped by the contemporary popular culture that accompanied their rise. We may not all love plastics but we have been living in a Plastics Age for a long time. As the millions of tons of waste plastics in the oceans degrade and enter the food chain, we are actually becoming plastics in a way that Andy Warhol could never have comprehended; we now have their synthetic chemistry in our bodies. We have to become globally responsible for how we produce, consume and dispose of plastics. We know that the material and its wider social and economic effects have made a profound impact on our lives and will continue to do so. Its place in the shifting world of popular culture is testimony to this fact.

BIBLIOGRAPHY

Albertine, Viv. 2014. *Clothes Music Boys*. London: Faber & Faber.

Anonymous. 1947. Plastics…a Way to a Better, more Carefree Life. *House Beautiful*, October.

———. 1957. How to Shop for Home Plastics. *Good Housekeeping*, July.

Banham, Reyner. 1976. The Style: 'Flimsy…Effeminate'? In *A Tonic to the Nation. The Festival of Britain 1951*, ed. Mary Banham and Bevis Hillier, 190–198. London: Thames and Hudson.

Barthes, Roland. 2000. Plastic. In *Mythologies*, trans. Annette Lavers, 97–99. London: Vintage. (Original French Edition 1957).

Bilbey, Diane. 2019. Britain Can Make It: the Dream of the Future. In *Britain Can Make It*, ed. Diane Bilbey, 17–28. London: Paul Holberton Publishing.

British Plastics Federation. n.d. https://www.bpf.co.uk//sustainability/Plastics_in_Transport.aspx. Accessed 20 April 2020.

Brook, Richard. 2017. Architecture Lecturer Helps 'Space Age' Plastic Classroom Win Listed Status. https://www2.mmu.ac.uk/news-and-events/news/story/6445/. Accessed 20 April 2020.

Bullivant, Lucy. 1986. Design for Better Living and the Public Response to Britain Can Make It. In *Did Britain Make It? British Design in Context 1946–86*, ed. Penny Sparke, 145–155. London: The Design Council.

Catterall, Claire. 1990. Perceptions of Plastics: A Study of Plastics in Britain, 1945–1956. In *The Plastics Age*, ed. Penny Sparke, 66–73. London: Victoria & Albert Museum.

Clarke, Allison J. 1999. *Tupperware, the Promise of Plastic in 1950s America*. Washington: Smithsonian Institution Press.

Cohn, Nik. 1989. Today There Are No Gentlemen. In *Ball the Wall*, 268–269. London: Picador.

Council of Industrial Design. 1946. *Britain Can Make It. Exhibition Catalogue*. London: CoID.

Darling, Elizabeth. n.d. Exhibiting Britain: Display and National Identity 1946–1967. https://vads.ac.uk/learning/designingbritain/html/bcmi_intro.html. Accessed 20 April 2020.

Freinkel, Susan. 2011. *Plastic: A Toxic Love Story*. Boston, NY: Houghton Mifflin Harcourt.

Galbraith, John Kenneth. 1958. *The Affluent Society*. New York: Houghton Mifflin.

Gossel, Peter, and Gabriel Leuthauser. 2005. *Architecture in the 20th Century*. Koln: Taschen.

Hebdige, Dick. 1979. *Subculture: The Meaning of Style*. London: Methuen.

Heimann, Jim. 2014. *Mid-Century Ads – Advertising from the Mad Men Era*. Koln: Taschen.

Hoggart, Richard. 1957. *The Uses of Literacy*. London: Pelican.

Howe, Sean. 2012. *Marvel Comics: The Untold Story*. New York: Harper Collins.

IKEA. 2018. *IKEA Catalogue*. IKEA.

Johnson, Alan. 2014. *This Boy*. London: Corgi.

Katz, Sylvia. 1984. *Classic Plastics: From Bakelite to High-Tech*. London: Thames & Hudson.

Meikle, Jeffrey L. 1990. Plastics in the American Machine Age 1920–1960. In *The Plastics Age*, ed. Penny Sparke, 40–53. London: Victoria & Albert Museum.

Moore, Kevin. 2000. *Museums and Popular Culture*. London: Cassell, Leicester University Press.

Morgan, Bill. 2010. *The Typewriter Is Holy*. New York: Free Press.

Mossman, Susan, ed. 1997. *Early Plastics: Perspectives 1850–1950*. London: Leicester University Press / Science Museum.

Ono, Yoko. 2010, September 19. Tweet. www.imaginepeace.com. Accessed 20 August 2015.

Picketty, Thomas. 2014. *Capital in the 21st Century*. Cambridge, MA: Belknap Press of Harvard University Press.

Savage, Jon. 2015. *1966: The Year the Decade Exploded*. London: Faber & Faber.

Song facts. n.d. https://www.songfacts.com/facts/jefferson-airplane/plastic-fantastic-lover. Accessed 20 April 2020.

Torgoff, Martin. 2001. *Can't Find My Way Home – America in the Great Stoned Age*. New York: Simon & Schuster.

V&A. n.d. Search the Collections. Daniel Weil. https://collections.vam.ac.uk/item/O85208/radio-in-a-bag-radio-weil-daniel/. Accessed 20 April 2020.

Violet, Ultra. 1988. *Famous for 15 Minutes, My Years with Andy Warhol*. New York: Avon Books.

Wasserman, John. 1966. *San Francisco Chronicle*, May 23.

Waugh, Evelyn. 1957. *The Ordeal of Gilbert Pinfold*. London: Chapman & Hall.

Weil, Daniel. 1986. Fluent in Plastics. In *Our Domestic Landscape*, 31. Cornerhouse: Manchester.

Weinstein, Dave. n.d. Plastic Fantastic Living. www.eichlernetwork.com/article/plastic-fantastic-living. Accessed 20 April 2020.

Westwood, Vivienne, and Ian Kelly. 2014. *Vivienne Westwood*. London: Picador.

Windsor, John. 1994. Identity Parades. In *Cultures of Collecting*, ed. John Elsner and Richard Cardinal, 49–67. Cambridge, MA: Harvard University Press.

Wright, Nicky. 2000. *The Classic Era of American Comics*. London: Carlton Books.

Yarsley, Victor E., and Edward G. Couzens. 1941. *Plastics*. London: Pelican.

CHAPTER 10

But They're Only Imitation…? Plastic Flowers That Can Disgust and Delight

Kirsten Hardie

This chapter considers the cultural, historical and commercial value of plastic flowers. It appears that, like some plastics artefacts such as doilies, rain bonnets and cutlery, plastic flowers may be regarded as inconsequential and anonymous, objects used but not necessarily scrutinised; their designs may not be readily attributed to a particular designer or manufacturer and their histories appear to be elusive. Nonetheless, artificial flowers made of plastics are ubiquitous and can be seen to proliferate within a range of private and public spaces across the globe. They bloom readily and serve a variety of functions across different cultures and countries. Yet, their appearance and very presence can be provocative. They can be perplexing and provoke harsh criticism. The artificiality of their form and material may be seen as problematic. Through a critical evaluation of plastic flowers as common objects, this study questions the inherent worth of plastic flowers whose existence, function and appeal is shaped largely by the very material of which they are made (Fig. 10.1).

K. Hardie (✉)
Arts University Bournemouth, Poole, UK
e-mail: khardie@aub.ac.uk

© The Author(s), under exclusive license to Springer Nature
Switzerland AG 2020
S. Lambert (ed.), *Provocative Plastics*,
https://doi.org/10.1007/978-3-030-55882-6_10

177

Fig. 10.1 Plastic flowers, manufacturer unknown. Date unknown. (Photo: K. Hardie, 2019)

Underpinned by key theoretical perspectives regarding material culture and notions of taste, the chapter examines the provocative nature of artificial flowers and examples made from plastics in particular. It considers how the plastic qualities of their being can position their place within hierarchical, social, cultural and design rankings. It discusses how they can rouse extreme responses: they can be abhorred and also be cherished; they can be dismissed as kitsch or embraced as things of beauty. The study examines the resilience of examples and explores their ability to appeal and endure despite accusations that they are worthless, cheap and nasty. The significant connotations that they carry, their exchange and symbolic value are considered. The displeasure and offence and charm and appeal of plastic flowers are debated. Why they are rejected or revered and the role that personal and popular taste plays in the evaluation of design in plastics is scrutinised. In particular, this chapter considers plastic flowers in relation to the concept of kitsch and notions of fakery (Bayley 2017; Kjellman-Chapin 2013).

The chapter firstly outlines what artificial flowers are, the different types of materials they are made from and the various contexts in which they exist. It then focuses on plastic flowers and highlights there historical development and popularity, notably in relation to consideration of the issues relating to plastics and perceptions of plastic flowers' worth. Whilst the term 'plastic flowers' is used, artificial flowers are made of many differ-ent plastics. Plastics such as polyester can create more fabric-like, lifelike flowers; however, it is often the hard, brittle plastic flowers that appear more unreal. In its consideration of the presence, popularity and provoca-tive nature of plastic flowers, notions of taste and different attitudes towards plastics and plastic flowers, it draws on the views of designers, key manufacturers, academics and professionals associated with design, horti-culture and floristry.

Artificial flowers and plastic flowers in particular, can appear to be a vexing area of design. The veracity of plastic flowers, their artificiality, copying of nature, and their synthetic form, is problematic. However, this chapter pres-ents an important and original reappraisal of plastic flowers and their worth, arguing that they have and continue to make, an important contribution to design and culture, even though they can disgust as well as delight.

ARTIFICIAL FLOWERS

Artificial flowers are designs created to simulate natural examples. The term 'artificial flowers' has traditionally been used as the generic term for all imitation flowers and does not necessarily relate to the individual specif-ics of the design or the materials used. Other terms used include faux flora, forever flowers and permanent botanicals. Increasingly, many contempo-rary flower designers and manufacturers strive to copy nature more accu-rately, intricate designs, applied finishes and coatings and the detailed use of materials increasingly present unreal flowers which appear surprisingly, and on occasion, convincingly real.

A variety of natural and synthetic materials has been used to create arti-ficial flowers, for example, paper, ceramic, shells, soap, sugar, glass, stone, wood, wax, feathers, leather, felt, latex, nylon, polyester, velvet, satin, rayon, cotton and silk. Notably the term '*silk* flowers' has become errone-ously the euphemism for artificial flowers in general. Polyester flowers, widely popular, are often incorrectly referred to as silk.

Importantly, a range of plastics, such as polyethylene and polyester, has shaped artificial flowers' mass production. Plastics' mouldable nature has enabled detailed shapes that have imitated nature effectively and also crudely. Whilst different plastics are used, so often all flowers are bunched together and identified as just 'plastic'.

The history of artificial flowers in general is not core to this study's focus; however, their long existence is acknowledged. Artificial flowers were produced by ancient Egyptians and 'over 1000 years ago in China for the emperor's wife who wanted flowers of all seasons in the palace all year round' (Simon Pykett, email to author, 24 May 2016) and 'In the mid-1700s, their popularity bloomed in France, and from there spread to England and America' (Museum of Early Trades and Crafts 2019). They were particularly popular in Victorian times where the artificial flower trade evidenced the manufacture of examples by homeworkers and the development of factories and training homes where child flower workers toiled.

Hand-crafted artificial flowers have existed as both commercial and home activities: a craft and a hobby, occasionally promoted as a 'profitable sideline' (Camm 1957, 23). The 1960 'Artificial Flowers' book claims 'though most things have a vogue, perfect and beautifully modelled flowers will always be appreciated' (Howard 1960, 5) and today, kits such as 'Forever Flowerz' by Craft Buddy Limited (2020), and origami guides and books continue to encourage their creation.

Artificial flowers are highly versatile. They are convenient to use, durable, long lasting, anti-allergy, colourful and easy to clean. They are used in a variety of ways, for example, millinery, decoration, gifts, retail displays, bridal bouquets, and so on. In hot countries, plastic flowers offer durability, for example, plastic Phuang malai (flower garlands) in Thailand operate as a colourful long-lasting good luck sign, and the traditional Mid-Autumn Festival in China sees bright plastics 'pig-in-a-basket' mooncakes' packaging adorned with vibrant flowers. A wide variety of flower-types are created and they may even be scented to satisfy a greater sense of the real. Their colours can be dramatic. Their purchase is perhaps an excusable indulgence or an acceptable and convenient alternative to their real counterpart.

Past encounters with artificial flowers or prevailing social attitudes may shape views and understanding of artificial blooms regardless of actual individual ownership of examples. They may be dismissed en masse despite the significant differences in the material-types used and the quality and craftsmanship of different designs and manufacturers who produce them. Whilst many artificial flowers may be relatively costly, they still may be derided as cheap and disgusting. For some, the alternative terms used to identify artificial flowers, such as faux flora or permanent botanicals, cannot disguise or excuse the fact the flowers are fake. Artificial flowers are stigmatised, as author Avi Steinberg (2017), writing about handmade fake flowers, commented:

When I profess my affection for fake flowers, I often feel as though I'm confessing a character flaw. They have, to say the least, a bad reputation. As decoration, they are considered tacky; as gifts, tactless. They are widely regarded as creepy and depressing.

Professor Jeffrey Meikle (email to author, 23 September 2019), American cultural historian and design historian, and author of the book *American Plastic: A Cultural History* (1995) stated:

> On the whole I find them depressing, except for the ones that are so well done that I don't notice they aren't real biological flowers. They're especially depressing when they're dusty or faded. They indicate people who have no time to give to maintaining the real thing. Occasionally I've been in a restaurant or office that has no natural light whatsoever, and plastic flowers in hanging baskets against the walls or on tables do relieve the gloom or impersonality.

Yet, today many flowers are of a high quality and perhaps are more amenable to reticent customers who are cautious of the artificial. For example, Peony™ Faux Flowers is 'the only artificial flower manufacturer to carry the prestigious and much respected Royal Horticultural Society Royal stamp of approval for the most botanical [*sic*] correct faux flowers worldwide' (Peony 2020), and the British Florist Association accredits, and labels accordingly, examples that they deem as excellent. Margaret Ashbourne (email to author, 5 January 2018), founder of Peony™ Faux Flowers, observes that the popularity of artificial flowers 'now is due to so many being indistinguishable from fresh flowers' and 'Pricing—although slightly more expensive, than fresh, the longevity of faux makes them vastly cheaper overall.' Thus, the issues and examples discussed evidence the various uses and attitudes towards artificial flowers and how their manufacture can aid their particular value and ensure their longevity.

PLASTIC FLOWERS

Plastic flowers can manifest intricate designs and detailed manufacturing processes. Some may have many components, for example, metal stems that fasten petals centrally and jointed petal sections that make detailed flower heads. Design patents reveal flowers' designs and production methods and demonstrate how skilled manufacturers copyright their designs in a competitive market. Today flowers are often hand assembled and individual components may be made by different manufacturers. The mass manufacture of artificial flowers is also a high-tech process. Flower components can be injection moulded, microwaves are used to dry dyed petals, and acrylic coatings are applied to protect flowers from damaging UV light. The detail of some flowers evidences how far nature is imitated, for example, rose stems may have prickly thorns and petals may sport resin drops to create the illusion of fresh dew.

Two misconceptions regarding plastic flowers are that they always fade and that they are everlasting. Whilst they can fade, as light and water can dilute their colours, increasingly many plastic flowers are resilient and colours remain strong. They can also die: some become brittle and disintegrate into tiny dry shards. Flower heads can smell, a distinct plasticky odour, their stability weakens, petals and leaves can tear and crack, and wire stems coated in plastics may split as the plastics deteriorate. The ugliness of plastics' decomposition can influence opinions. Thus, sensitive to consumers' desires for long-lasting faux flowers, many designs use materials that can ensure a bloom's longevity to secure its grace.

HISTORICAL OVERVIEW OF PLASTIC FLOWERS

The sprouting and growth of plastic flowers has been shaped by the creation, development and use of particular plastics. The identification of the birth of the plastic flower appears to be elusive however. Cellulose nitrate (or celluloid as it is known in the US) fashioned flower jewellery popular in the 1930s and 1940s and Larson (1993, 28) reported celluloid and vinyl flowers were made in Japan in 1955 and 1956. Many plastic flowers have been made of 'polyethylene and polyvinyl' using 'injection, moulding, and extrusion' processes (Larson 1993, 28). Boyle (1962) suggested that plastic flowers were created in the 1950s commenting, either as 'a daisy made in Italy' or 'a poinsettia' for the high street retailer, Woolworth. Meikle's important study of plastics (1995, 255) highlights their availability in 1951 and Lo (2017) reported: 'that the mass production of plastic flowers became feasible in 1954 thanks to 7-parts plastic molds for [a] rose developed by John Corelli of New York (proprietor of Corham Artificial Flowers), Alfred Jean Fristot of France and Lino Bosco of Italy.'

The success and failure of plastic flowers continues to be shaped and determined largely by their symbiotic relationship and association with plastics. Where plastics develop and are popular, plastic flowers grow and where plastics are troublesome or disliked, the flowers are tainted accordingly. For example, in the late 1940s and early 1950s, consumers were wary of plastics as their poor experiences of products made from low-quality plastics damaged their trust in the material (Wahlberg 1999, 34). Plastics were viewed as 'inferior ersatz' (Meikle 1995, 125). In response, the Society of the Plastics Industry, the American trade association, attempted to ensure that plastics products were labelled to inform consumers. Whilst discussions ensued regarding the introduction of plastics' quality standards (Wahlberg 1999, 13), such measures were not implemented and consumers' and the industry's concerns regarding the quality of plastics remained. However, when 'the success of plastics during the 1950s was inarguably staggering' (Wahlberg 1999, 5), plastic flowers basked in plastics' glory.

The popularity of plastic flowers bloomed in the 1950s and manufacturers in a number of countries, including France, Germany, America and Japan, produced a diverse range and a prolific number; as Meikle (1995, 255) observed, one American company was producing '200 varieties' of plastic flowers. Key plastic flower manufacturers included Corbosco, Italy; the Corham Artificial Flower Company, America; the California Artificial Flower Co., America; and Frisot's Prestige Floral, France. These manufacturers' flowers were of high quality. By the late 1950s, plastic flowers increased in popularity and production, and 'Hong Kong emerged as the world's leading producer' (Lo 2017).

In America, consumers increasingly embraced the readily available flowers. In 1961, 'more than $112 million worth of plastic flowers and plants were sold in the United States, while only $16 million worth were sold in the rest of the world' (Breskin Publications, Inc., 1962 cited in Larson 1993, 29). In 1964, the *New York Times* reported that 'The palpable evidence of polyethylene plant proliferation has grown' (Kohn 1964). Increasingly, private and public spaces featured plastic flowers although the quality of the designs and the plastics used were variable. Concerns regarding the quality and veracity of plastics continued; 'plastic's [negative] reputation as an imitative substitute clung stubbornly' (Wahlberg 1999, 75) and plastic flowers were tarnished by association.

In the UK, plastic flowers' popularity increased also and in the early 1960s, they were a successful marketing tool used by brands as premiums (free gifts) to lure consumers. Manufacturers, Procter and Gamble, offered a free rose with a box of Daz (Mullin 2014, 165) and Lever Brothers offered a daffodil to promote Omo. As Lever Brothers Chairman Len

Hardy (cited in Bowen 1995) reported, 'Some people laughed at us, [...]
But many others lived in the middle of industrial cities, and would hardly
ever see flowers. The promotion also had the great attraction that customers would come back time and again to collect a bunch'. It was reported
that '8 million roses were given away [...]' by Procter and Gamble (Mullin
2014, 165).

Today a wealth of artificial flowers and plants are readily available.
Examples are manufactured by a myriad of companies in numerous countries including Hong Kong, South Korea, Taiwan, Europe and Thailand
and the majority are produced in China (Global Artificial Flower Market
2019). The financial value of artificial flowers is huge, and sales are significant, and the array of flowers available is sizeable.

Interestingly, in a competitive global market with numerous manufacturers, many have remained anonymous. As Busch (2007) claimed:
'Secrecy, indeed deception, is intrinsic to artificial flowers (none bears a
fixed label, naturally enough) and details of their manufacture are hard to
discover'. However, some manufacturers have identified their products
and have secured paper labels to present makers' details and country of
origin. For example, the 'Plastic Flower' brand, Hong Kong and the
Giftcraft Original (c. 1980, country unknown) readily identified their
products. Notably, Prestige Floral placed a label on the stems of their
moulded polyethylene flowers that proudly stated, 'Les Creations D 'Art
A. Frisot, Made in France'. Interestingly, and importantly, they also
included their name, '© PRESTIGE FLORAL' and a year date, stamped
into the long plastic-covered wire stem and on the underside of some of
the leaves. Yet, whilst many of today's faux flowers are readily labelled,
many flowers continue to remain anonymous.

Taste and Attitudes: Plastics and Flowers

Throughout history, a plethora of design artefacts has been acknowledged
and evaluated; their form and function have been considered and their
histories have been documented and examples have been collected by
museums. Numerous artefacts are identified as design classics: 'industrially
manufactured objects of aesthetic value and timeless quality [...] definitive
models of lasting influence and enduring significance' (Phaidon editors
2006) and these inform and shape our understanding of design and culture. However, some artefacts such as disposable cutlery, personal shopping trolleys and shower caps, although omnipresent and long serving as

convenient functional items, have not received such formal attention. In a consideration of the documentation and examination of the history of design, Julier (2008, ix) comments 'more ordinary, everyday stuff of design [...] [is] largely overlooked and underestimated'. Such appears to be the fate of plastic flowers. Rarely evident in museum collections, they have appeared not to warrant academic scrutiny. The reasons why some artefacts appear not to have been considered in any depth and why an evaluation of their use and worth appears neglected, may be in part because they are relatively inexpensive and are mass produced; they may be discarded readily and replaced easily. It may be because they replicate or imitate nature and are thus inauthentic. It may be just because they are made of plastics. Whilst Barthes (2000, 97) spoke of plastics as a 'miraculous substance', attitudes towards plastics are ever-changing and as Fisher (2015, 127) observed, 'through their history their identification with an inauthentic "synthetic" origin has meant that plastics have gained a negative image'. Plastics are highly versatile and extensively used to serve our daily lives but views on plastics are polarised. Newport (1976, 1) observed, 'Objects made of plastic tend to be rejected by all traditional forms of recognition, in a way other important materials are not' and this seemingly continues and there appears to be a problem whether it is the plastics or the object that is most important (Rabolini 1983, 16). Thus the appeal and value of a plastic flower may be shaped by it being 'plastic' regardless of the excellence of its design and manufacture, or the type and quality of the plastics used. With today's acute environmental concerns and the grave evidence of plastics' detrimental impact, increasingly superfluous plastics designs are abhorred. As artist Arjan Van Arendonk (email to author, 9 April 2020), whose vibrant creations feature flowers, stated, 'They should be forbidden, like all other plastics for that matter'; and as Buranyi (2018) reports: 'Plastic is everywhere, and suddenly we have decided that is a very bad thing. Until recently, plastic enjoyed a sort of anonymity in ubiquity: we were so thoroughly surrounded that we hardly noticed it. [...] Decades after it became part of the fabric of our lives, a worldwide revolt against plastic is under way'.

As such, the existence of the plastic artificial flower is questioned yet interestingly new plastic flowers have emerged in response to environmental concerns. For example, the contemporary British designer Sarah Turner creates 'everlasting' flowers from upcycled plastics bottles (email to author, 30 March 2016).

Thus, plastic flowers are objects that have associated meanings. The power of objects to appeal to individuals and to communicate specific meanings has been discussed extensively (e.g. Forty 1986; Baudrillard 2005; Barthes 1957; Sudjic 2008; Bayley 2012; Boradkar 2010). Our use of design objects is a daily affair and our choice of, and connection with, them can vary significantly. Our relationship with objects can go beyond their use value; as Sudjic (2008, 49) observed, '[...] there is something to understand about objects beyond the obvious issues of function and purpose.' Objects can 'transmit meanings' (Forty 1986, 6) and may represent and be given or associated with certain values, social and cultural positions and meanings, and these can change. For example, the rose is a popular flower type made of plastics. The rose as an established cultural symbol of love and beauty has certain connotations whatever material it may be made from. The gift of a rose can communicate thanks, affection, passion, life and perhaps eternal love. The natural rose may be discarded when wilted or kept and cherished in its faded form. Like its real counterpart, the plastic rose can serve as a symbol of emotion too, it has an exchange value, and it may endure and retain its form and bright colour. The extent to which true love is represented by a fake rose and the sincerity of the giver may be questioned however. The plastic flower, whilst a symbol of love, may also be abhorred and it may, by association, reveal or lead to the questioning of the owner's sensibility and taste. The artificiality of a fake rose can compromise its meaning and value and indeed its owner.

Objects may appeal to us because of their form, their status, or cost, or for many other reasons and we fill our homes with a myriad of items. We can form strong emotional attachments with objects: we may love specific objects. Objects can be deeply personal and reflect individual tastes and indeed may be seen to reflect people themselves, as Csikszentmihalyi and Rochberg-Halton (1981, 16) stated, 'people are what they attend to, what they cherish, what they use.' Objects can be aesthetically appealing and as Harper (2018, 61) observed, 'An object's aesthetic expression, and sensuous qualities are decisive in determining the extent to which it is perceived as being attractive—and, in part, as being useful—by the subject.'

However, different subjective attitudes and tastes prevail. 'Tastes (i.e. manifested preferences) are the practical affirmation of an inevitable difference' (Bourdieu 1984, 56) and thus it can be observed that particular objects are positioned in a hierarchy of taste determined by such differences. Objects are judged and categorised in relation to highly subjective criteria. But personal reactions regarding something being good or bad taste also operate on a personal level, as Vercelloni (2017, 16) remarked: 'Taste is seen as the subjective ability to discern things intuitively, a gut reaction to an external stimulus, be this beautiful or ugly, sophisticated or obvious, refined or vulgar, authentic or fake.' However, to judge an object, such as a plastic flower, solely by its appearance can negate its inherent meaning and value, as Bayley (1992, 11) stated, 'Taste is not so much about what things look like, as much as about the ideas that gave rise to them.'

Often plastic flowers are deemed as kitsch; judgement is passed that they are trivial, cheap, imitations, sentimental, poor quality and of bad taste. Referred to as 'tasteless mass rubbish' (Bayley 2012, 135), 'kitsch' can be seen as a damning term. However, 'Although frequently dismissed as facile, lowbrow, or one-off, throwaway aesthetics, kitsch is surprisingly mobile and complex' (Kjellman-Chapin 2013, ix) and to reject an object as simply kitsch denies consideration of the object's fuller value and appeal. Thus, perhaps plastic flowers may be considered to be what Kjellman-Chapin (2013, ix) refers to as 'guilty pleasures'; we know they are fake and they may not be great quality, but as French poet Charles Baudelaire commented, 'What is so intoxicating about bad taste is the aristocratic pleasure in being displeased' (Ward 1991, 17). When Dr Susan Mossman, chair of the Plastics Historical Society, and formerly senior curator of Materials Science, Science Museum, London, was asked, 'Do artificial flowers (whether fabric or plastic) appeal to you?' Her response was 'Only if they are well made or dreadfully kitsch' (email to author, 26 February 2018). Similarly, designer Wayne Hemingway's response was 'no [they don't appeal] other than having a limited kitsch value and something to laugh at if particularly sad examples' (email to author, 4 January 2018). Thus, people can delight in plastic flowers as objects whose appeal and function primarily are aesthetic and fun, classic kitsch hallmarks, whilst finding some pleasure in the poor design and the poor quality of particularly lacking designs.

Consumers can enjoy objects that are playful and humorous and plastic flowers as decorative features can delight. For example, the c.1950s–1960s inflated footstool of plastics made in Japan, often referred to as a terrarium or diorama stool, has plastic flowers encased in its transparent bulbous belly. The flowers enhance the design's kitsch appeal. Likewise, a colourful bunch of injection moulded polyethylene tulips, made by Tobar in 2000, with small fairy lights in each flower delight when illuminated. Whilst functional, offering a colourful glow, overall it is the plastics quality of the flowers that shines through. Plastic flowers can thus be aesthetically appealing and operate as an arresting novelty. Their artificiality may be the essence of their being and their key appeal. The obviously plastic flower, its plastics components (stem, leaves, stamen, etc.) that suggest a real flower rather than intricately copying it, the bright-coloured petals that do not match a real flower's colour, texture or look, can be pleasing. We may knowingly delight in its jolly fakery and can display it unashamedly.

Through the consideration of objects in relation to notions of taste, it can be observed that particular designs acquire meanings and associations that are shaped by cultural and social attitudes and experiences. Some objects, such as plastic flowers, may sit precariously in relation to judgements of taste as views differ and decisions regarding their worth can place them either as nonsensical items or as valuable expressions of emotion.

Plastic Flowers and the Home

Encounters with artificial flowers can occur in a variety of places such as hotel lobbies, shopping malls and individuals' homes. An arrangement of faux flowers can be a distinct interior design feature, cheery decoration, a splash of colour, and can create a particular mood. Whether displayed as single stems or sprays in vases, as swags or major centrepieces, faux flowers are long lasting, low maintenance, handy alternatives to real flowers (Heath 2017). The artificial blooms are displayed to be enjoyed.

Artificial flowers in an individual's home evidence their choice in how they wish to decorate, shape and present their personal space. Home as a 'private sphere' (Miller 2001, 1) offers a place for self-expression and personal preferences. In the homes of others, where individuals' personal tastes and private possessions differ or jar with our own, the sight of fake flowers can be provocative and memorable. As Tilley (cited in Petridou 2001, 88) observed, 'Places are contexts for human experiences, constructed in movement, memory, encounter and association' and thus an artificial flower arrangement in the home may become associated with a particular person, event or moment in time. An old, dusty, faded plastic rose bouquet may appear unloved but it may not be readily dismissed or discarded as it may evoke important personal memories and its appearance transcends judgement as its inherent value is far greater than its physical state. As Miller (2001, 83) commented, 'Things embody relations and memory'; thus, we may forgive the pitiful look of aged faux flowers when they are charged with personal association and specific values and meanings.

However, memory and nostalgia may not save artificial flowers from scorn and indeed their exclusion within the home. When questioned why can artificial flowers and especially plastics versions be ridiculed and regarded as poor design, why do you think this is? Meikle (email to author, 9 September 2019) replied: 'Not only because they are such pathetically unconvincing imitations, but because they have no practical function other than to relieve their owners of having to maintain real living plants, which do have small functions: releasing a bit of oxygen, reminding us inside of the larger world outside of which we are a part, and testifying to the dedication of their owners in maintaining them.'

This underscores that individuals' perceptions of flowers' fakery, the designs' synthetic materials and their mimicry of nature, can position artificial flowers, and particularly plastic flowers, as unconvincing imitations, copies that cannot, and should not, take the place of authentic flowers.

Yet fake flowers appear ever more in public urban spaces, as seasonal decoration in shopping areas including flower arches framing shop facades. As a marketing approach, retailer Apple showcased gigantic artificial flowers as window displays in Selfridges, London, in 2015, to launch the new Apple watch. The huge flowers demanded attention. By comparison, a rich display of plastic flowers in hanging baskets adorns the frontage of the Beehive public house in Liverpool city centre (UK). The flowers serve as an important decorative function and the displays are changed twice a year so different flowers offer colour in winter. The flowers attract attention and are cost-effective compared to real flowers. Likewise, vibrant plastic flowers hang in baskets positioned proudly along the platforms of some UK railway stations. This suggests a nod towards a wider acceptance and reappraisal of plastic flowers as functional forms that add colour and life to our more mundane places. However, UK florist Helena Horton (2019) comments, 'Not only does [*sic*] the plastic plants confuse bees and butterflies in search of nectar, it encourages the manufacture of needless plastic products.' Whilst synthetic flowers appeal, sensitivities and negative attitudes exist. For example, gardening expert and former Vice President of the Royal Horticultural Society, Alan Titchmarsh MBE, commented, 'Plastic flowers were never very good, but have now been replaced with brilliantly made silk flowers which are extremely good. I have several arrangements!' (email to author, 9 June 2018).

Interestingly, fake flowers are used as expressions of emotion and devotion. Depending upon culture, religion and personal preferences, artificial flowers are placed upon graves or at commemorative monuments, for example, they are regularly positioned at the foot of the statue of pop star Billy Fury (created by Tom Murphy) in Liverpool, UK. The flowers are a sign of remembrance evidencing a certain piety. The symbolism of particular flowers, whether a single rose, a small posy or a wreath, can emphasise specific associations. Flowers are a mark of respect for a loved one, a potent symbol of grief and admiration, a decorative addition. For example, plastic coffin flowers have been used in place of real wreaths. Their bright and sturdy design offers durability and convenience. However, the existence of fake flowers in a graveyard can prove to be particularly incongruous and provocative. They may draw significant scorn from those who view their use as poor taste and a breach of graveside decorum. As Meikle (email to author, 9 September 2019) commented: 'Regarding the importance of plastic flowers today, I suppose they fill insignificant cultural and economic niches. Otherwise they have no importance at all. Maybe I should add that they are especially depressing on graves.' For some, only real flowers are appropriate and the ban of artificial flowers in some cemeteries confirms their expulsion.

The stigma that artificial flowers have long suffered appears to continue. Tensions remain regarding the aesthetic value of plastic flowers, as Busch (2007) remarked: 'Yet, despite their vaulting verisimilitude, "faux flowers" cant alter (that queasy genteelism is still common) retain cheap associations redolent of shoddy, late-20th-century Taiwanese mass manufacture'. However, the popularity of faux flowers continues. Whilst plastic flowers can wilt or fade, they appear to weather the criticisms that too often relegate them as tasteless and worthless.

Perhaps plastic flowers are more accepted, or more readily forgiven, at certain seasonal times when displayed as cheery decorations. Christmas plastic holly wreaths, imitation poinsettias and faux fir trees entice consumers. They can mimic nature quite closely. A particular bold use of artificial flowers was created by the St. Anthony Museum, Lisbon, Portugal, in June 2019. The museum used a mass of artificial flowers to celebrate the annual month-long feast of St. Anthony. The artist, Suzana Barros, was invited to develop their idea of a celebratory wall of flowers that would endure in bright sunshine. The museum's external façade packed with artificial bright blossoms presented an arresting sight that created a complete wall of colour over eight metres high, providing a dynamic and memorable welcome to the museum's entrance. The flowers proved most popular as Museum Coordinator Pedro Teotónio Pereira (email to author, 22 August 2019) explained: 'Throughout June, at any time of day or night, there is always someone photographing or taking selfies on the flower wall. The number of visitors to the Museum and the Museum store has increased considerably [...]. Many people ask to leave the flower wall all year round, but for us it's important to keep it as a seasonal installation, related to the feast of St. Anthony, not losing the surprise effect, which is the key element of this project.'

Conclusion

This study presents a fresh consideration of artificial flowers through examination of the value of plastic flowers as design examples that have largely remained somewhat neglected and undocumented. It has discussed how plastic flowers are artefacts that provoke significant reactions that can disgust and delight. Whilst plastic flowers continue to be derided and associated with bad taste, cheapness and poor quality, a long entrenched stigma, this chapter has demonstrated that they also remain popular and widely used across a variety of contexts. This study's research explicitly shows that plastic flowers appear in many places and that their quality and life-like appearance has improved significantly in recent years. For some, a plastic flower is to be cherished, a thing of beauty that is pleasing to display; for others, the plastic flower is regarded as poor show. However, attitudes regarding colourful plastic flowers are not just black and white. People's perceptions vary dramatically: tastes are volatile and plastic flowers can be particularly provocative. Ultimately, artificial flowers, and notably the plastic flower, remain a matter of personal taste.

Artificial flowers and plastic flowers in particular may continue to carry negative associations. They may be embraced, tolerated or rejected depending upon their look or the context of their use. It is concluded that tensions continue to surround the use and evaluation of plastic flowers owing to their charged connotations and reputations that are shaped by the fact that they are artificial and that they mimic the real ... and often just because they are made of plastics.

This chapter makes a new contribution to the established literature relating to artificial flowers, plastics and design. Ultimately, it is recommended that further research into plastic flowers is pursued to more fully recognise their complexities and importance as valuable design objects and to reappraise their enduring worth as blooming marvellous designs.

Bibliography

Barthes, R. 1957. *Mythologies*. London: Paladin.
———. 2000. Plastics. In *Mythologies*. London: Vintage.
Baudrillard, J. 2005. *The System of Objects*. London: Verso.
Bayley, Stephen. 1992. *Taste: The Secret Meaning of Things*. London: Pantheon Books.
———. 2012. *Ugly: The Aesthetics of Everything*. London: Goodman Fiell.
———. 2017. Is There Such a Thing as Good Taste? *The Telegraph*. http://www.telegraph.co.uk/luxury/design/design-critic-stephen-bayley-thing-good-taste/. Accessed 12 September 2018.
Boradkar, Prasad. 2010. *Designing Things: A Critical Introduction to the Culture of Things*. Oxford: Berg.
Bourdieu, P. 1984. *Distinction: A Social Critique of the Judgement of Taste*. Cambridge: Mass.: Harvard University Press.
Bowen, David. 1995. When Plastic Flowers Ruled The Earth. *The Independent*, October. https://www.independent.co.uk/news/business/when-plastic-flowers-ruled-the-earth-1579961.html. Accessed 18 March 2020.
Boyle, Hal. 1962. A Rose Is a Rose Is a Rose Or Is It? Step Right Up, Folks, and See If You Can Tell Difference Between Real and Realistic Flowers. June 5. *St. Louis Post*, 37. https://www.newspapers.com/newspage/141611230/. Accessed 4 May 2018.
Buranyi, Stephen. 2018. The Plastic Backlash: What's Behind Our Sudden Rage – And Will It Make a Difference? *The Guardian*, November 18. https://www.theguardian.com/environment/2018/nov/13/the-plastic-backlash-whats-behind-our-sudden-rage-and-will-it-make-a-difference. Accessed 19 September 2019.
Busch, Simon. 2007. So Good, They're Almost Real. *Financial Times*, July 13. https://www.ft.com/content/6e10f78c-3017-11dc-a68f-0000779fd2ac. Accessed 12 September 2018.
Camm, A.F.J. 1957. Artificial Flowers. *Practical Home Money Maker* October: 23–24.
Csikszentmihalyi, Mihaly, and Eugene Rochberg-Halton. 1981. *The Meaning of Things: Domestic Symbols and the Self*. Cambridge: Cambridge University Press.
Fisher, Tom. 2015. Fashioning Plastics. In *The Social Life of Materials*, ed. Adam Drazia and Susanne Kuchler, 119–135. London: Bloomsbury.
Forty, Adrian. 1986. *Objects of Desire: Design and Society Since 1750*. London: Thames and Hudson.
Global Artificial Flower Market 2019: Market Key Players, Growth, Trends, Revenue, Share and Demands Research Report and Forecast to 2025. 2019. https://www.theexpresswire.com/pressrelease/Global-Artificial-Flower-Market-2019-Market-Key-Players-Growth-Trends-Revenue-Share-and-Demands-Research-Report-and-Forecast-to-2025_10335355?mod=article_inline. Accessed 18 March 2020.

Harper, Kristine H. 2018. *Aesthetic Sustainability: Product Design and Sustainable Usage*. Abingdon: Routledge.

Heath, Olivia. 2017. How to Style Artificial Flowers and Plants at Home. *House Beautiful*, April 27. https://www.housebeautiful.com/uk/decorate/display/g149/faux-artificial-flowers-plants-styling-tips/. Accessed 3 December 2019.

Horton, Helena. 2019. Flower Arches Trend Is Damaging the Environment Because Many Cheat with Plastic Blooms, Say Top Florists. *The Telegraph*. https://www.telegraph.co.uk/news/2019/07/05/flower-arches-trend-damaging-environment-many-cheat-plastic/. Accessed 3 December 2019.

Howard, M.D. 1960. *Artificial Flowers*. London: Mills and Boon.

Julier, Guy. 2008. *The Culture of Design*. 2nd ed. London: Sage.

Kjellman-Chapin, Monica, ed. 2013. *Kitsch: History, Theory, Practice*. Newcastle: Cambridge Scholars Publishing.

Kohn, Sherwood. 1964. The Flowering of Fake Flowers. *New York Times*, August 23. https://www.nytimes.com/1964/08/23/archives/the-flowering-of-fake-flowers.html.

Larson, Roy A. 1993. Impact of Plastics in the Floriculture Industry. *HortTechnology* January/March 3 (1): 28–34.

Lo, York. 2017. The Forgotten Jewish American Tycoons Behind the HK Plastic Flowers Industry. *The Industrial History of Hong Kong Group*, December 18. https://industrialhistoryhk.org/the-forgotten-jewish-american-tycoons-behind-the-hk-plastic-flowers-industry/. Accessed 18 January 2018.

Meikle, Jeffrey L. 1995. *American Plastic: A Cultural History*. New Brunswick, NJ: Rutgers University Press.

Miller, Daniel, ed. 2001. *Home Possessions: Material Culture Behind Closed Doors*. Oxford: Berg.

Mullin, Rody. 2014. *Promotional Marketing: How to Create, Implement and Integrate Campaigns That Really Work*. London: Kogan Page.

Museum of Early Trades and Crafts. 2019. Artificial Flowers. http://www.metc.org/artificial flowers/. Accessed 17 May 2019.

Newport, Roger. 1976. *Plastic Antiques: An Exhibition of Plastic Products from the 1850s to the 1950s. British Industrial Plastics Limited*. Wolverhampton: Wolverhampton Polytechnic.

Peony. 2020. About Peony Fake Flowers. https://www.peony.co.uk/about-julie. Accessed 18 March 2017.

Petridou, Elia. 2001. The Taste of Home. In *Home Possessions: Material Culture Behind Closed Doors*, ed. Daniel Miller, 87–104. Oxford: Berg.

Phaidon editors. 2006. *Phaidon Design Classics*. London: Phaidon.

Rabolini, Anna. 1983. In Gli Anni di Plastic, by ed. Pasquale Alferj and Francesca Cernia. Milan: Electra.

Steinberg, Avi. 2017. Letter of Recommendation: Fake Flowers. *The New York Times*, April 12. https://www.nytimes.com/2017/04/12/magazine/letter-of-recommendation-fake-flowers.html. Accessed 24 September 2018.

194   K. HARDIE

Sudjic, Deyan. 2008. *The Language of Things*. London: Penguin.
Vercelloni, Luca. 2017. *The Invention of Taste: A Cultural Account of Desire, Delight and Disgust in Fashion, Food and Art*. Trans. Kate Singleton. London: Bloomsbury.
Wahlberg, Holly. 1999. *1950s Plastic Design: Everyday Elegance*. Atglen, PA: Schiffer Publishing.
Ward, Peter. 1991. *Kitsch in Sync: A Consumer's Guide to Bad Taste*. London: Plexus Publishing.

CHAPTER 11

Ambiguous Artificiality: The Presentation and Perception of Viscose Fibres and Fabrics in Norway in the 1930s

Tone Rasch

Plastics led to fundamental changes in global material culture during the twentieth century. The substantive breakthrough arose from the 1950s onwards when major amounts of plastics, developed during World War II, were released on to the commercial market for the first time. This impacted also on textile production leading to the synthetic revolution. However, in the case of textiles, this development was heralded between the wars, when, before the introduction of oil-based fibres, like nylon and polyester, semi-synthetic materials became popular (Coleman 2003, 933).

These semi-synthetic fibres were made from pulped natural materials high in cellulose rather than fossil fuels. They were invented in the 1880s and called artificial silk on account of their shiny silk-like texture. Then, in the 1930s, they became available with more matt surfaces under different trade names, like viscose, the American name rayon and the German name Zellwolle, referring to the mix of cellulose and wool, emphasising the

T. Rasch (✉)
Norwegian Museum of Science and Technology, Oslo, Norway
e-mail: tone.rasch@tekniskmuseum.no

© The Author(s), under exclusive license to Springer Nature   195
Switzerland AG 2020
S. Lambert (ed.), *Provocative Plastics*,
https://doi.org/10.1007/978-3-030-55882-6_11

natural textile connection. In Norwegian it was called 'cellull'. During the inter-war years, more than one hundred companies entered the artificial fibre field worldwide (Woodings 2001, 11).

Viscose was put into production all over the industrialised world including in smaller European countries like Norway, which is the focus of this chapter. Today, the names viscose and rayon are still in use. Neither the viscose textiles nor the synthetic fibres have fully substituted for natural fibres but have acted as supplements to meet the increasing demand for textiles during the century. They should not, however, be viewed solely as substitutes. New inventions like stretch and breathing fabrics would not have been possible without synthetic fibres (Meikle 1995, 171–172).

Viscose fibres hold an ambiguous position in between natural and synthetic fibres. The basic material is cellulose, made from plant pulp, treated with chemicals like sodium hydroxide and carbon disulphide. There is great resistance to oil-based textile fibres but a sense that, because viscose fibres are plant based, they are closer to nature and therefore kinder materials, more likely to biodegrade, although this is not always the case (Museum of Design in Plastics). Certainly, they have the merit of being made from renewable feedstock which gives them an advantage over production and consumption of synthetic oil-based fibres today. The large environmental organisation in Norway 'Framtiden i våre hender' (Future in our hands) categorises textiles in two main groups, as natural and synthetic fibres, based on the ecological footprint the textiles leave behind. Rayon and viscose are grouped with natural fibres, like plant and animal-based fibres and are recommended for use in preference to oil-based synthetic fibres, which leave behind micro plastics from their production, wearing, washing and wasting (Helle n.d.).

The natural feel of viscose fabrics has also increased their popularity. However, their production can sometimes be bad for the environment. The website goodonyou.eco, which deals with sustainability in fashion, explains: 'much of the viscose on the market today is manufactured cheaply using energy, water and chemically-intensive processes that have devastating impacts on workers, local communities and the environment' (Robertson n.d.). It is still possible to produce sustainable viscose fibres but at a higher cost, which raises the question of balance between economic manufacture versus protection of the environment.

The blurred borders between the natural and artificial character of viscose still apply to the materiality of the fibres today. However, it is not only a question about their materiality. It relates also to their social and cultural value. Viscose textiles float between different aspects of technology and society. The impact of viscose fibres and fabrics has changed as new inventions and new ways of use within different fields of consumption have become available. Close to the outbreak of World War II, Europe was experiencing difficult economic times. Manufacture of synthetic textiles was one solution to keep the factories working.

Chemical production of the fibre itself was not the only challenge. It's processing also required special machinery. When first used, the filamented viscose fibres were cut into shorter lengths, called stapled fibres. These were spun, knitted and woven on traditional machinery, mainly designated for the cotton industry. The new fibre became in that way an integrated part of the traditional textile industry (Singleton 1997, 88). In this text, I explore what these viscose fabrics looked like, how they were used and when they became widely familiar.

Mass media played an essential part in the growth of synthetic fabrics. In the years between the wars, newspapers and weekly magazines exerted a growing influence as introducers of new ideas. The design historian, Grace Lees-Maffei, has examined perceptions of designed objects in the paper 'The Production-Consumption Paradigm', published in *Journal of Design History* in 2009: 'the role of channels such as television, magazines, corporate literature, advice literature and so on in mediating between producers and consumers' (351). She demonstrates that the object can be understood through the role of the producer as well as the consumer, and by the way it is represented and discussed in the media. Viscose fabrics and clothes are objects that have been thus mediated and, thereby, been made available and desirable to consumers.

Norwegian Society in the 1930s

To understand the development and the position of viscose in the Norwegian textile industry, it is necessary to look closer at the society of the 1930s. This period of transition in Europe between the two world wars was characterised by a process of modernisation, economic decline and social unrest. The situation in Norway was no exception. Norway had not participated in World War I but was, nonetheless, financially and politically affected by it. Relations were especially close with Great Britain and Germany. These countries were Norway's main trading partners and where many of the more highly educated had studied. Norway also had close economic and cultural links with the other Nordic neighbouring countries.

Norway was sparsely populated, and the economy dominated by primary industries like farming and fishing. However, with the arrival of the industrial revolution in the middle of the nineteenth century, increasingly the population also worked in factories, including textile factories. Traditionally, spinning and weaving mills imported technical knowledge as well as machinery and cotton from Great Britain (Bruland 1989). Then, at the beginning of the twentieth century, Norwegian industry entered a new phase. The rich water resources were transformed by hydro-electrical power and electrochemical companies were set up.

The census from 1930 provides a review of the Norwegian population and the different industries. The population was at that time less than three million inhabitants. Over a third of the gainfully employed worked in farming and fishing (*Folketellingen i Norge* 1934, 2). About 770,000 were employed in manufacturing. The main breadwinner in the family was usually the man, although the number of workers was almost equally divided between men and women. The textile industry was dominated by spinning and weaving mills for wool, cotton and knitting industries. Almost 11,600 women and 5600 men were employed in the textile industry, totalling 17,700 employees. The clothing industry, including shoe factories, milliners and manufacturers of other accessories, was bigger, having been restructured from small industries and craft set-ups to factory operations. According to the census, almost 50,000 were employed in these industries (*Folketellingen i Norge* 1934, 51).

Usually, the census was carried out every ten years but because of World War II, the next census was not until 1946. By then 19,220 people worked in the textile industry, made up of about 12,000 women and 7000 men (*Folketellingen i Norge* 1954, 51). The wool industry was the largest. The viscose and the artificial silk industry employed only about 700 people: 500 men and 200 women. The ready-to-wear industry had by comparison 46,000 employers, one-third men and two-thirds women. Other important industries in Norway at this time were the metal industries with 90,000 employees, and the food industry with 40,000 employees.

The Norwegian textile and clothing industry depended on international fashion, which foresaw the trends. The fashion industry, as it was presented in the press and through clothing stores, strengthened the national clothing industry. Traditionally, fashion was taken from Paris and given a national flavour (Rasch 2020). The national impact on fashion increased towards World War II, both politically and for economic reasons. It became important that the clothes were made in Norway and expressed

Norwegian values. Architecture drew inspiration from previous building practices. National costumes became more popular and were used for festivities (Haugen 2010). The construction of national identity was mainly based on folk art from vernacular pre-modern society, as it had been idealised since the end of the nineteenth century. This trend continued throughout the inter-war years although society itself was rapidly adapting to increasing industrialisation. There was an obvious discrepancy between the preindustrial ideals and the reality of industrialisation.

VISCOSE PRODUCTION BEFORE 1936

Cellulose nitrate fibres were invented by the French chemist Hilaire de Chardonnet and exhibited as 'artificial silk' at the world exhibition in Paris in 1889. This material was the precursor to viscose fibre (Handley 1999, 19). The chemical industry continued to improve the production of man-made fibres using cellulose and chemical additives. Dyeing and finishing had for years used chemical processes within the mechanical textile industry to produce textiles of natural fibres. Production of cellulose fibre further increased these contact points leading in the first decades of the twentieth century to the development of a range of viscose fibres in Europe, the United States and Japan. Courtaulds built a large factory in England in 1905. The American Viscose Company was established in 1911 and in 1915 the first Japanese plant Yonezawa was built (Singleton 1997, 88) (Fig. 11.1).

Germany had several leading chemical plants producing dyestuff from the nineteenth century onwards. These companies extended production to synthetic fibres. In the same way as synthetic dyes took the place of natural dyes, artificial fibres took the place of natural textile fibres. Viscose was a good substitute for silk, but ironically its silky surface was a limitation where clothing was concerned as it meant it could not fully substitute for cotton and wool, the most common materials in clothes. However, this drawback was overcome in the 1920s. The German company Köln-Rottweiler Pulverfabriken AG launched a cotton-like viscose fabric Vistra, and soon after Wollstra, a mix of Vistra and wool (Völkel 1938, 35–36, 82–84). Agfa, BASF, Hoechst and Bayer had all been leading German manufacturers of dyestuffs and chemicals for the textile industry. These companies fused under the name I.G. Farben, developing new products for the textile industry as well as for the growing armament industry in the

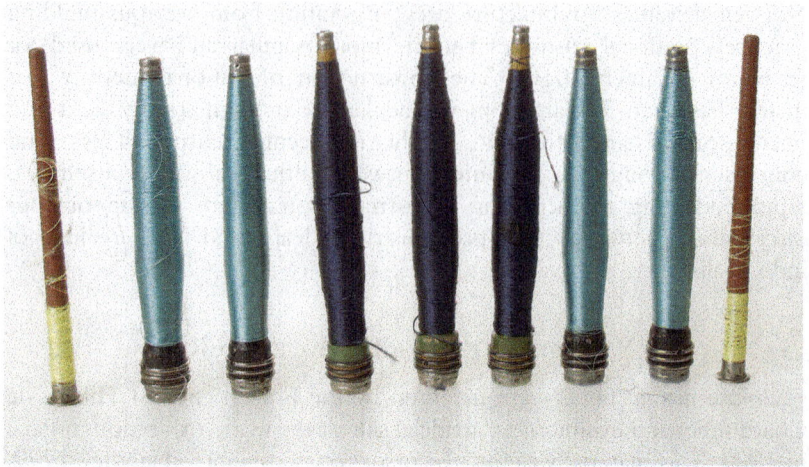

Fig. 11.1   Nine coils for viscose yarns, used by Hjula circa 1950. (Photo: Finn Larsen, Norwegian Museum of Science and Technology)

inter-war years. Improvements to the viscose fibres continued within this group of companies.

In Norway, the textile and chemical industries were too small to take part in the race to produce new artificial fibres. The weaving mills and knitwear industry depended on importation of both technology and fashion from the leading European industrial centres. However, new trends quickly became evident and it did not take long for Norwegian textile factories to start experimenting with viscose. The first known tests with dying artificial silk in Norway were done by the Nydalens Compagnie in Oslo in about 1890, but due to problems with the fabric's durability, it was not put into production at this time (Hjula Væveri Archive, Norwegian Museum of Science and Technology, A-1018, Xbc Prøvebøker for lager, Nydalens Compagnie, L0002 Diverse prøver). Then, in 1910, the J. Jacobsen tricotagefabrik in Larvik became the first Norwegian factory to use artificial silk commercially (Grieg 1948, 2: 314). Nonetheless, Norwegian viscose production remained fragmented until 1924 when the Nye Høie Fabrikker (New Høie factories), outside Kristiansand, started producing artificial silk fabrics, branded *Høie-silke* and *Høie-Damask* (Benestvedt 1950, 75). Artificial silk tapestries woven by the mill were exhibited at the national industrial exhibition in Bergen in 1928.

Business records from the textile industry often contain sample books which provide unique information about the textiles being produced, information that would otherwise be lost. Usually, these samples are well preserved, protected from the light within books in the archives. The records from the cotton mill Solberg Spinderi outside Drammen, about fifty kilometres west of Oslo, are preserved at the factory, still existing as a trade company today. The company started to spin cotton as early as 1818 and around one hundred years later expanded with a weaving department. A sample book in the archive named *Kulørt Picquet 1923 Kunstsilke* presents their first experiments with artificial silk in 1926. The designs were fabrics with a mix of lustreless cotton yarn and shiny viscose, called Bemberg-Adler-Silke made by the cuproammonium process, produced in Germany (Völkel 1938, 65; Woodings 2001, 99–100).

Hjula Væveri (Hjula Weaving Mill), situated in Oslo, was a leading Norwegian textile factory in the 1930s. Substantial records relating to it are preserved at Norsk Teknisk Museum (Norwegian Museum of Science and Technology), including a large collection of sample books showing the production of the cotton weaving and print departments. One undated sample book, probably from the second half of the 1920s, contains samples made mainly from 100% cotton but also includes two fabric types named *Kval.B* and *Kval.C* with sub numbers, which are interwoven with viscose patterns. The fabrics from Hjula are similar to the fabrics in the aforementioned Solberg sample book, showing that these early examples were part of an ongoing trend.

The account books from Hjula confirm the use of viscose as a regular part of this Norwegian company's production of fabrics in the second half of the 1920s. In August 1926, the mill created a special account for silk yarn which probably meant artificial silk (Hjula Archive, A-1018, A-Fabriksreskontrobøker, L0029 Fabriksrescontro 1917–1932). Agfa and Snia Viscosa were the main suppliers. Agfa was one of Germany's pioneers in viscose production. The factory in Berlin originally produced dyestuff. But from 1918 onwards, the factory had two parallel divisions, one for photo and film and one for viscose production. The research laboratory at the plant gave Agfa a leading position within the development of semi-synthetic fibres (Löhnert and Gill 2000, 126–127). The successful lustreless qualities Vistra and Wollstra, referred to earlier, were made at this factory (Völkel 1938, 82, 84). Additionally, the Hjula's general yarn account also documents the factory's procurement of viscose yarn from other German spinning mills that had changed from spinning cotton to viscose during these years. Kulmbacher Spinnerei in Bayern, Leipziger Baumwollspinnerei and Industrie-Gesellschafte für Schappe in Basel were the most important traders for Hjula from the end of the 1920s onwards (Hjula Archive, A-1018, A-Fabriksreskontrobøker, L0029 Fabriksrescontro 1917–1932). Viscose was also bought from Snia Viscosa, the largest Italian viscose company, which produced viscose from the 1920s onwards. The semi-synthetic textile production in Italy grew rapidly with plants in North-Italian cities and headquarters in Milan (Lupano and Vaccari 2009, 266).

Hjula had traded with foreign textile and chemical companies for decades. Thus, this business with the German viscose producers was a continuation of normal practice. However, that with the Italian manufacturer was something new. Hjula must have searched for suppliers to satisfy the company's high demand for the new viscose yarn. The practice continued until the mid-1930s: viscose yarn and woven fabrics were imported by the Norwegian mill and sold as Norwegian goods.

Promotion of Norwegian Fabrics in Norwegian Magazines

Information about the new synthetic fabrics was conveyed to consumers, usually women, by weekly magazines. The role of the fashion magazines in the hard years before and during World War II has been described as 'a fantasy space to escape' (Guenther 2004, 369). Women's magazines worked the same way in Norway. Since the end of the nineteenth century, Norwegian journalists had reported home from fashion centres like Paris, Vienna or Copenhagen. They told fairy tales about the glorious fashionistas of Paris, giving the homebound readers a sense of luxury. However, they also provided a mix of sensible education and entertainment, encompassing cooking, knitting, family life and reports about celebrities of the time. The popularity of the magazines grew during the inter-war years, and new printing techniques allowed them to be richly illustrated with photographs. Editorials included reports on fashion and interior design as well as descriptions of the changes in fashion and textile production. The synthetic fibres also appeared in advertisements for Norwegian consumer products.

Two monthly magazines, *Vi selv og våre hjem (Ourselves and Our Homes)* and *Hus og have (House and Garden)*, reveal how the fibres were portrayed. Usually, *Vi selv og våre hjem* presented the viscose fabrics as reasonable and sensible alternatives to clothes made with natural fibres affordable by the average reader. The article 'Feriegarderoben planlegges' ('The Holiday Wardrobe Is Planned') is an example of this. The target audience was clearly upper middle-class housewives, planning their summer holiday by the seaside for the whole family, including the maid. The different synthetic materials: artificial silk, Vistra and the elastic yarn Lestex, a composite yarn with a rubber core, used for swimwear, were described in the text alongside the varied clothes the family would need for social occasions and leisure (Blich 1935, May). It was also possible to buy the fabrics and patterns presented in the magazine (Anonymous 1935, May). Mostly the fabrics were mentioned by their brand name, like *Bjørg*, *Magna* and *Drafn* from Solberg Spinderi, without reference to their particular fibres. However, in the case of Hjula's fabric *Vistracheviott* and the fabrics from Nye Høie the viscose fabrics are described as an integrated part of the wardrobe. In the case of Nye Høie's in an advertisement for fabrics for women's clothes, it was mentioned that two of them contained Wollstra (Advertisement 1935, April) (Fig. 11.2).

Although their reception was largely favourable, there was also a critical voice. In an article entitled 'Norske stoffer' ('Norwegian fabrics'), new collections of textiles produced by the Norwegian industry were presented with photographs of the fabrics and their manufacturers. The author was Knut Greve, one of the leaders of the arts and crafts organisation, *Foreningen Brukskunst*, which aimed to encourage good taste among consumers. He was engaged in social work to improve the living conditions of working-class people through the new design movement, organising exhibitions and writing books about affordable, hygienic and tasteful interiors. In the article, he identified two directions within the country's textile design: Norwegian and European. He described the new viscose fabrics thus: 'The one [tendency] as in the modern Wollstra or Vistra fabrics and the artificial silks is a distinct European look, which does not differ significantly from other countries' textiles in terms of appearance'. His preference was for gingham: 'This is an old, Norwegian specialty that has been improved year by year according to taste as well as quality. There is no doubt that we are the leading European producer and that gingham now means summer, sun, holiday' (Greve 1935). Gingham, although made from imported cotton and dyes, had been produced by the Norwegian textile industry since the 1850s. At that time, it was a foreign element within traditional Norwegian textile production but eighty years later it had become Norwegian in a way that was acceptable to Greve, as a defender of Norwegian design.

Fig. 11.2   Models, wearing dresses made by fabrics woven at Hjula. The photo was published in the woman's magazine *Hus og have*, June 1936. (Photo: Moreno Tollaas, Norwegian Museum of Science and Technology)

There were also reports from the more extravagant Parisian fashion sites in which it was often taken for granted that the clothes featured were made of expensive fabrics. Only exceptionally were specific fibres mentioned but then only as silk and wool. It was as if the new fibres were not used within haute couture. An example of this understanding is provided by the journalist Ferdinand Finne writing about the Parisian designer Jeanne Lanvin in *Hus og have* (Finne 1935a, January). The clothes were described in detail with their colours and fabric types, for example, velvet, crêpe, and georgette, with no explanations whether the fibres were synthetic or natural.

However, changes were on the way. A month or two later, he wrote about glassy fabrics designed by Elsa Schiaparelli as 'the new glass-fabric used for a tunic, shoes and a handbag, all transparent. If you want to be ahead of your time, buy Schiaparelli's glass dress or glass tunic that Paris smiled sceptically on last autumn' (Finne 1935b, April), a reference to the brittle and transparent spun synthetic, Rhodophane (Handley 1999, 27), she developed with the French textile company, Colcombet. Schiaparelli pioneered the innovative use of synthetic materials within haute couture, predicting the synthetic revolution that took place after World War II. Thus, Finne mediated the novelty of artificial materials within the fashion industry to Norwegian readers, supported by fashion photography, produced by the Schiaparelli fashion house itself.

STARTING VISCOSE PRODUCTION IN NORWAY

Viscose production in Norway was initiated as a response to increased demand for viscose textiles in the Norwegian market and as a means of helping to eliminate unemployment. It was thus a matter of national significance written about in newspapers, which at that time were owned by political parties. They were targeted mainly at men and their focus was not so much on fashion but rather on economy, politics and production.

The first Norwegian artificial silk factory, Kunstsilkefabrikken, was established in 1936 at Notodden, a small town in Telemark southwest of Oslo. Both industrialists and some of the political parties were involved. The existing pulp industry provided a firm foundation based on the electrochemical as much as on the mechanical industry. Notodden was one of the main production sites of the electrochemical company Norsk Hydro, producing fertilisers from 1905 onwards. In the inter-war years, the company, as well as the local economy, were severely stricken by structural changes related to the worldwide economic crisis. The average unemployment in Notodden was currently more than 30%, about 10% higher than the average unemployment in Norway (Orning 2015). There was political agreement about the need to strengthen industry to fight the economic crisis on a national level. One proposal was to increase the domestic production of consumer goods by establishing new factories using new technology. The director of Norsk Hydro, Axel Aubert, had already in 1930 suggested expanding the plant's production in Notodden by establishing a viscose factory (Aubert 1930, 33). The company even had close connections to I.G. Farben through fertiliser production methods. Although these plans were never carried out, they contributed positively to the establishment of the new factory there some years later.

Kunstsilkefabrikken was built in co-operation with Kunstseidenwerke Küttner in Pirna in Saxony, Germany. At the start, it had about 120 employers and produced both the glossy artificial silk and the stapled cotton-like artificial material (Aftenposten 1937a, August 10). These yarns, which were used to make a range of different products such as linings, football jerseys, knitting yarn and table clothes, were sold internally to the Norwegian textile industry. Also, in 1936, the weaving mill Tele silkeveveri was founded in the neighbourhood of the spinning mill, enabling local production of fabrics from the viscose yarn.

A key player in establishing Kunstsilkefabrikken was Arne Bergsvik (1892–1968) (Digitalarkivet), a small-scale businessman who had lived in Germany in the 1920s when he had sympathised with the anti-communist movements (Bergsvik 1943). Back in Norway, he had no political mission until he became mayor for the Nazi party Nasjonal Samling (National Rally) during the war in Notodden. In the 1930s, he argued convincingly for the establishment of a Norwegian viscose plant (Bergsvik 1937, March 17) and when in 1935 a parliamentary committee was appointed to give an account of reasons to support viscose production in Norway, he became the committee secretary (Lykke and Havardstad 1935). It was the first time the Norwegian state intervened in the establishment of an industrial factory.

The press was invited to visit the newly opened factory in August 1937 with the result that journalists from leading newspapers commented on the trip. Overall, the opinions expressed were positive, regardless of their political views. They were welcomed by the factory's management: the chairman and lawyer, Blom, the general manager, Bergsvik, and the engineer, Wahlberg. The funding of the factory was a topic of special interest: a journalist of the Labour Party newspaper *Arbeiderbladet* wrote a short note about the visit, emphasising the trouble encountered in raising the necessary capital, which was solved by state guarantee, initiated precisely by the Labour Party (1937b, August 10). Following discussion of the financial situation, the production processes were explained. Cellulose was delivered by two of the largest cellulose factories in Norway,

Borregaard and Saugbrugsforeningen. Journalists emphasised the significance of Norwegian businesses starting domestic production of viscose making importation of foreign-made textiles unnecessary. A journalist from the liberal newspaper *Tidens Tegn* (*The Signs of the Times*) ended his text praising the possibilities of the new fibre: 'It is rather curious to think, as we nowadays marvel at all the beautiful and well-dressed ladies who promenade along our thoroughfares, that the vast majority be clad in fabrics stemming from the spruce forests, from their suntan-hued stockings all the way up to those smart, diminutive hats that adorn their well-groomed coiffures' (1937, August 10).

Dagbladet, another liberal newspaper covered the visit (1937, August 10) with a text signed by Sir Anthony, the pseudonym of Anton Beinset, a central pressman representing the liberal party, Venstre (Sommerfeldt 1936). He described the journey through the experiences and statements of participants. One was Ole Colbjørnsen, a Labour Party politician who wrote for *Arbeiderbladet*. He had played an important role in the development of the party's financial plan in the 1930s (Andersen and Yttrii 1997, 130–131), contributing to a three-year plan to strengthen Norway's economy. Kunstsilkefabrikken was part of this. Sir Anthony's other participant in the report was a nameless female journalist who expressed her lack of technical competence in a parodic manner, repeating 'Oh! I wonder' and 'Oh! How strange', and: 'It couldn't occur to me to wear anything but trees on my body!' These comments can be seen as a way of expressing the general lack of technical knowledge about viscose production of the average consumer.

Efforts to strengthen Norwegian industry took place on several levels. Oslo Håndverks-og Industriforening (the Oslo organisation for crafts and industry) celebrated its centenary in 1938 with the exhibition called *Vi kan* (*We can*) (Gramm 2008). It provided an important showcase for industry and was a popular success, with more than one million visitors. Three halls exhibited products made solely by factories located in Oslo, one of which was devoted to textiles. Fashion shows of evening dresses, everyday dresses and suits for middle-class women complemented the displays. A review in the women's magazine, *Alle Kvinners Blad* (*All Women's Magazine*),[1] carried the excited headline: 'We can also dress Norwegian—and still be as chic as any *parisienne*' (1938, 27). The text emphasised the clothes' connections to different lifestyles and the captions mentioned the names of the fabrics and the fibres from which the dresses depicted were made. The term 'cellull' was used as frequently as cotton.

The fabrics cited had been woven by the leading viscose clothing fabric factories in Oslo: the Hjula Væveri and Nydalens Compagnie.

Hjula had just invested in printing machinery and used the opportunity to launch a completely new collection of printed furnishing fabrics. They were presented in an article in *Alle Kvinners Blad*. The name of the most mentioned collection was *Oseberg*, deliberately recalling the famous Viking burial mound with a ship inside, that a few years earlier had acquired its own museum in Oslo. The patterns were, however, not at all inspired by the Viking style but rather in the popular vernacular neo-baroque style. They were described as 'sophisticated' with 'hand-printed charm', and 'best of all: the low price' (Jeanne 1938). The article was supported by advertisements in the weekly press, promoting the new decorative printed textiles (Advertisement 1938, July). Interestingly, neither the article nor the advertisements mentioned viscose (Fig. 11.3).

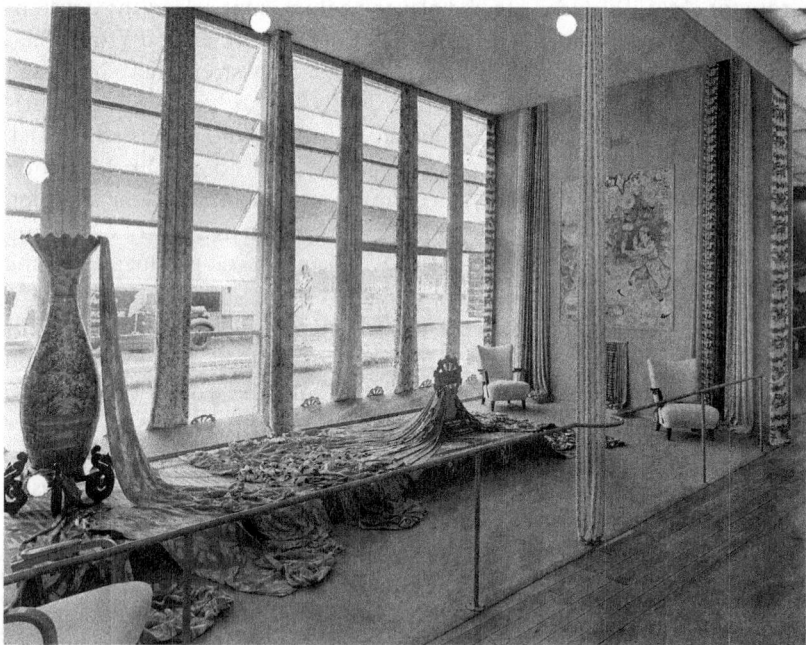

Fig. 11.3 Hjula Væveri's exhibition at *Vi kan*, Oslo 1938, exhibiting printed furnishing textiles in viscose. (Photo: Unknown, Norwegian Museum of Science and Technology)

Viscose Production in the Archive

Two books of patterns in the Hjula archive provide opportunities to examine the new printed designs more closely. They have an unusually large format and are both labelled *Trykkvareprøver 1939–1941* and contain about 40 design titles spread over 360 pages (Hjula Archive, A-1018 X-Stoff-og mønsterprøver, L0006 Trykkvareprøver 1939–40). The pattern samples have scarcely any information other than their titles and numbers. However, they reveal very different characteristics. *Oseberg*, already referred to, and similar pattern types, made from artificial silk are rough and intended for curtains and other furnishing textiles. They were made from yarn from Kunstsilkefabrikken as the mill bought yarn from the factory in 1938 and 1939 (Hjula Archive, A-1018 Da-Inngående korrespondanse L-0354-0360 Diverse inngående korrespondanse T 1938–1947, letters from Tele Silkeveveri). Others were for thin dress fabrics, some simply called *Cellull-Musselin*, while others have titles like *Ekeberg*, which is a district in Oslo and *Camping*, referring to summer holiday activities. The designs' origins are not mentioned in the sample books (Fig. 11.4).

Fig. 11.4 Samples of the furnishing fabric *Oseberg*, silkscreen on viscose, from the sample book *Trykkvareprøver 1939–41* from the Hjula archive. (Photo: Håkon Bergseth, Norwegian Museum of Science and Technology)

The wide range of designs within the sample books follows the trend set by Hjula's collections of woven fabrics from the 1890s onwards. Factories usually had both permanent staff designers and bought designs from domestic and European design offices (Boydell 1995, 29). However, Hjula did not have in-house designers and had bought designs from abroad for a long time (Rasch 2003, 92). The mill's letter archive from the end of the 1930s is partially preserved and the travel account contains letters from the advertising manager, Rasmus Rivedal, to the management, describing a journey in early summer 1938. His first stop was Paris, where he visited the design office of J. Claude Frères from whom Hjula had bought designs since the 1880s. He also visited Galeries Lafayette and other department stores, noticing printed dress fabrics either in cotton, artificial silk or silk (Hjula Archive, A-1018, Da-Inngående korrespondanse L-0343 Diverse inngående korrespondanse R 1938–1947, letter 25 May1938). After Paris, Rivedal travelled to Prague, then to Reichenberg, today known as Liberec. There, he visited the textile mill Vereinigte Färbereien to discuss 'the new collection of Cellull-musslin' (Hjula Archive, A-1018, Da-Inngående korrespondanse L-0343 Diverse inngående korrespondanse R 1938–1947, letter 31 May 1938) (Fig. 11.5).

It seems likely that the collection of thin and printed cellull-muslin found in the Hjula pattern books are from the collection bought on Rivedal's journey. There were, however, political and military circumstances that could have made doing business difficult. In September 1938, Hitler annexed Bohemia in the Czech Republic. The German occupation did not lead to material destruction of the textile industry but to a gradual aryanizing of the trade, originally owned by Jews (Hlaváčková 2000, 9). The war situation made it difficult to import textiles from the area to Norway. A letter in the Hjula archive from De norske Tekstilfabrikers Hovedforening (the Norwegian textile manufacturers' employer organisation) reveals the chaotic situation of trying to sort out where the different factories were located, whether in Germany or Czechoslovakia (Hjula Archive, A-1018 Da-Inngående korrespondanse L-0354 Diverse inngående korrespondanse T 1938–1947, Letter 18 November 1938). A last letter about patterns produced in Czechoslovakia was dated in December the same year. Prices for cellull-muslin and flannels were mentioned but not confirmed. The letter was addressed to Rivedal, but the sender is unknown (Hujla Archive, A-1018 Da-Inngående

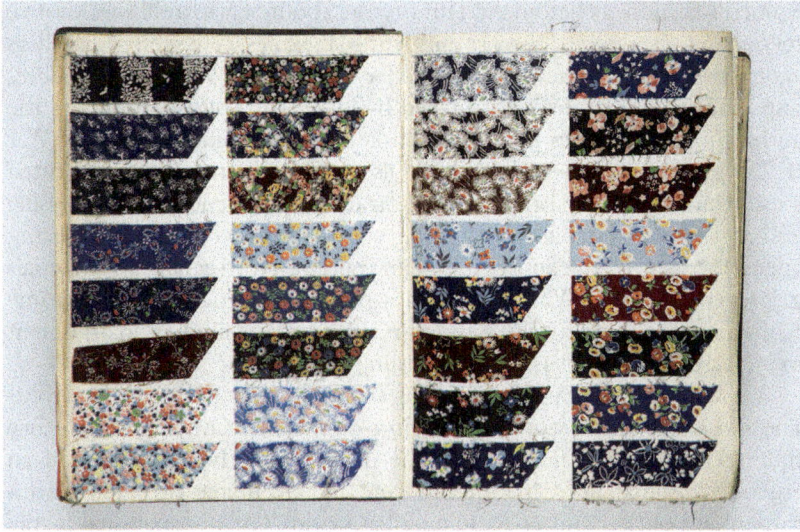

Fig. 11.5 Samples of the dress fabric *Cellull-Musselin* from the sample book Trykkvareprøver 1939–41, printed textiles on viscose, from the Hjula archive. (Photo: Håkon Bergseth, Norwegian Museum of Science and Technology)

korrespondanse L-0354 Diverse inngående korrespondanse T 1938–1947, Letter 18 November 1938). The situation, with a war approaching, may explain why these designs became a short-lived part of the Hjula collection. In September 1939, World War II broke out in much of Europe, and in April the following year, Norway was occupied by Nazi Germany and the textile industry had to adapt to new conditions.

Conclusion

The business archive from the textile mill Hjula Væveri has contributed to a deeper insight into what went on behind the scenes in terms of the development of Norwegian viscose as a transnational technology. The inter-war period can be divided into different phases, due to different attitudes to viscose fibres. The first half of the 1930s was dominated by the production of artificial silk, either as silky women's wear; elegant, but less expensive than natural silk, or as decorative effects in the production of mainly cotton fabrics. Journalists wrote about viscose being a part of women's fashion, as Ferdinand Finne wrote about Schiaparelli's innovative cellophane-like dresses. Viscose and artificial textiles were seen primarily as luxury products. However, in the latter half of the 1930s, viscose fibres responded rather to demands fuelled by increasing consumption and the underlying problems in the global markets that culminated in World War II. The popular cotton- and wool-like viscose fibres increasingly substituted for cotton and wool fabrics. Viscose production became a political issue, leading to the establishment of a state-sponsored factory reducing unemployment in an exposed area and utilising Norwegian raw materials to prevent imports. What had been a private clothing business became economically and politically significant. Fashion was used as an argument to increase production as well as consumption.

Was viscose provocative? How was artificiality portrayed and received? The ambiguity can be viewed as the fibre on the one hand imitating luxury and offering fashion novelty. On the other hand, the fibre satisfied an urgent economic need in offering new and profitable products into a hardening political and economic global world. The transnational textile industry depended on trade and technology transfer. The 1930s were no exception despite a pronounced protectionism. The early synthetic fibres based on cellulose rather than oil were during these years a welcome supplement to ensure the continuity of production. Environmental issues were less discussed. They were a matter for the future. At this time fashion, economy and technology were unified to utilise Norwegian natural resources, thereby strengthening the country's industry.

NOTE

1. The magazine *Hus og* have changed name to *Alle Kvinners Blad* in 1938, published weekly instead of monthly.

BIBLIOGRAPHY

A.O. 1937. Fabrikken på Notodden som spinner silke uten assistanse av silkeormen. *Tidens tegn*, August 10.
Advertisement. 1938. Sommerens sensasjon! Oseberg. *Vi selv og våre hjem*, July.
Advertisement. 1938. Hygge og fest. *Vi selv og våre hjem*, December.
Advertisement for the designs 'Cotelé' and 'Homespun'. 1935. *Hus og have*, April.
Andersen, Ketil Gjølme, and Gunnar Yttri. 1997. *Et forsøk verdt*. Oslo: Universitetsforlaget.

Anonymous. 1935. Norske sommerklær for sol og luft og såpevask. In *Vi selv og våre hjem*, May.

―――. 1937a. Eventyret om den norske gran som blir til klær. *Aftenposten*, August 10.

―――. 1937b. Kunstsilkefabrikken i drift. *Arbeiderbladet*, August 10.

―――. 1938. Vi kan klæ oss norsk. *Alle Kvinners Blad, 27.*

Anthony, Sir (pseudonym for Anton Beinset). 1937. Deres unevnelige er laget av – . *Dagbladet*, August 10.

Anton Beinset. i Norsk biografisk leksikon på snl.no. https://nbl.snl.no/Anton_Beinset. Accessed 8 May 2020.

Aubert, Axel. 1930. *Den Elektrokjemiske Industri i Norge og dens Fremtidsutsikter.* Oslo: Emil Mostue Trykkeri.

Benestvedt, Olav. 1950. *Høie Fabrikker 1850–1950.* Oslo: Tanum.

Bergsvik, Arne. 1937. Kunstsilkefabrikasjon i Norge? *Aftenposten*, March 17.

―――. 1943. Den eneste redning. *Tromsø*, December 15.

Blich, Milly. 1935. Feriegarderoben planlegges. In *Vi selv og våre hjem*, June.

Boydell, Christine. 1995. Free-lance Textile Design in the 1930s: An Improving Prospect? *Journal of Design History* 8 (1): 29.

Bruland, Kristine. 1989. *British Technology and European Industrialization.* New York: Cambridge University Press.

Coleman, Donald. 2003. Man-made Fibres before 1945. In *The Cambridge History of Western Textiles*, ed. David Trevor Jenkins, vol. 2, 933–947. Cambridge: Cambridge University Press.

Digitalarkivet. https://www.digitalarkivet.no/view/387/pc00000001881362. Accessed 4 May 2020.

Finne, Ferdinand. 1935a. Madame Lanvin. In *Hus og Have*, January.

―――. 1935b. Litt av hvert for våren og sommeren. *Hus og Have*, April.

Folketellingen i Norge 1 desember 1930. 6: Folkemengden fordelt efter livsstilling. 1934. Oslo: Det Statistiske Centralbyrå, https://urn.nb.no/URN:NBN:no-nb_digibok_2011021006016. Accessed 14 August 2020,

Folketellingen i Norge 3. desember 1946, VI: Yrkesstatistikk. 1954. Oslo: H. Aschehoug, 48–71. https://urn.nb.no/URN:NBN:no-nb_digibok_2011112308086. Accessed 14 August 2020.

Gramm, Geir Olav. 2008. Oslo som det moderne – Vi kan-utstillingen 1938. *Byminner, tidsskrift for Oslo museum* 3: 20.

Greve, Knut. 1935. Norske stoffer. In *Vi selv og våre hjem*, May.

Grieg, Sigurd. 1948. *Norsk tekstil.* Oslo: Tanum.

Guenther, Lisa. 2004. *Nazi Chic?* Oxford: Berg, Oxford.

Handley, Susannah. 1999. *Nylon. The Man-made Fashion Revolution.* Baltimore: Johns Hopkins University Press.

Haugen, Bjørn Sverre Hol. 2010. The Concept of National Dress in the Nordic Countries. In *Berg Encyclopedia of World Dress and Fashion Vol. 8 West Europe*, ed. Lise Skov, 18–21. Oxford: Berg.

Helle, K.E. Slik unngår du klær av plast. https://www.framtiden.no/gronne-tips/klar/slik-unngar-du-klar-av-plast.html. Accessed 3 April 2020.

Hlaváčková, Konstantina. 2000. *Czech Fashion 1940–1970*. Prague: Olympia.

Jeanne. 1938. Vi bor norsk, billig og vakkert. *Alle kvinners blad*, 26.

Lees-Maffei, Grace. 2009. The Production-Consumption-Mediation Paradigm. *Journal of Design History* 22 (4): 351–376.

Löhnert, Peter, and Manfred Gill. 2000. The Relationship of I.G. Farben's Agfa Filmfabrik Wolfen to its Jewish Scientists and to Scientists Married to Jews, 1933–1939. In *The German Chemical Industry in the Twentieth Century*, ed. John E. Lesch, 126–127. Dordrecht: John E. Kluwer Academic Publisher.

Lupano, Mario, and Alessandra Vaccari. 2009. *Fashion at the Time of Fascism*. Bologna: Damiani.

Lykke, Ivar, and Torvald Havardstad. 1935. Innst. S nr. 151 Tilråding på finans- og tollnemnda um trygd for eventuelt lån upptil kr. 500 000 til ein kunstsilkefabrikk i Notodden. *Stortingsforhandlinger (ib. utg.)*. 1935. *Vol. 84 Nr. 6a*: 305–312. https://urn.nb.no/URN:NBN:no-nb_digistorting_1935_part6_vol-a. Accessed 14 August 2020.

Meikle, Jeffrey. 1995. *American Plastics*. New Brunswick: Rutgers University Press.

Museum of Design in Plastics. Biodegradable Plant-Based Plastics. https://www.modip.ac.uk/projects/symbiosis/industry-interactions/symphony-environmental/biodegradable-plant-based-plastics. Accessed 3 April 2020.

Orning, H.J. 2015. Norges historie vekst, verdenskrig og depresjon, 1905–1939. Store norske leksikon. https://snl.no/Norges_historie. Accessed 3 April 2020.

Rasch, Tone. 2003. Blåtøy, kjoletøy og flanell. In *Volund Norsk Teknisk Museums årbok*, 73–106.

———. 2020. National Versus International. Norwegian Fashion Photography, the Clothing Industry and Women's Magazines. In *Fashioned in the North, Nordic Histories, Agents, and Images of Fashion Photography*, ed. Anna Dahlgren, 163–186. Stockholm: Nordic Academic Press.

Robertson, L. Material Guide: Is Viscose Really Better for the Environment? https://goodonyou.eco/material-guide-viscose-really-better-environment/. Accessed 3 April 2020.

Singleton, John. 1997. *The World Textile Industry. Competitive Advantage in world Industry*. London: Routledge.

Sommerfeldt, W.P. 1936. *Forfattermerker i norske aviser og tidsskrifter 1931–1935*. Oslo: Fabritius.

Völkel, Ernst. 1938. *Kunstseiden-und Zellwollarten*. Leipzig: Dr. Max Jänecke Verlagsbuchhandlung.

Woodings, Calvin. 2001. *Regenerated Viscose Fibres*. Cambridge: The Textile Institute, Woodhead Publishing Ltd.

CHAPTER 12

Plastics' Canonisation: Aspects of Value in the Camberwell Inner London Education Authority Collection

Maria Georgaki

The aim of this chapter is to use the example of a specific set of plastics items made in England to trace how approaches to value in plastics shifted during the 1950s and up to the late 1960s. The value of these artefacts will be considered within, and beyond, educational contexts. I will investigate plastics' position as quintessentially modern materials, and their subsequently ambivalent semantics. The focus will be on historic examples where plastics products were exalted for embodying the tenets of mid-century 'good design' and will address the value conflicts that this position engendered. Discussion will examine how an alignment with the aesthetics of modernism attempted to influence the perception of plastics as desirable materials and how this canonisation was problematised by the taste turn which gave rise to pop design in the late 1960s.

The artefacts in question are the plastics found in the Camberwell Inner London Education Authority (ILEA Collection) (henceforth 'the Collection'), currently one of the teaching collections of the University of

M. Georgaki (✉)
University of the Arts London, London, UK

S. Lambert (ed.), *Provocative Plastics*,
https://doi.org/10.1007/978-3-030-55882-6_12

the Arts London. The Collection was first assembled in 1951, under the titles of the Experiment in Design Appreciation and later, the Circulating Design Scheme. Emphasis on taste formation as a concern of the Collection is evident from the scheme's early title while in its next reincarnation, the aspect of circulation denotes the desire to disseminate this message.

The present study builds on the existing literature of travelling educational collections and the social history of plastics. In particular, it identifies the role of the objects in question as vehicles for ideas about production and consumption within the context of British secondary education. The chapter positions the Collection within the discourse about the impact of the 1951 Festival of Britain and connects it to projects which sprung from it. By discussing the involvement of local authority and state-funded design institutions, it demonstrates how the post-war 'good design' movement in Britain infiltrated London schools and presents an example of close association between schools and the design establishment in the 1950s and 1960s. Beyond the field of circulating collections, it extends the literature of critiques on modernism by examining how plastics which had won design accolades were received in the commercial sphere.

Locating Value in the Camberwell ILEA Collection

Between 1952 and 1976, the Circulating Design Scheme (later the Collection) was accessed through a formal lending programme, administered by the London County Council (later the Inner London Education Authority—ILEA). Objects were mounted inside purpose-built cabinets that were loaned to London secondary schools on a term-by-term basis. Displays were already arranged by the organisers. After transportation and remounting at the school, they revealed museum-style cases with objects and panels of explanatory texts.

The Collection as a resource brings inherent value to archive and museum education. Crucially for understanding its pedagogical value, and by extension, the pedagogical value of the plastics items within it, the Collection's present title recognises its origin as a scheme managed by the Inner London Education Authority, alongside the fact that it is now housed at Camberwell College of Arts.[1] Although it was only in 1990 that Camberwell College of Arts acquired the Collection, all the objects had been made and collected from the early 1950s to the early 1970s. Despite being initially rather larger, some items were deaccessioned due to severe damage; today, the Collection comprises about 6000 individual pieces, including multiples of the same design. There are in total about 200 plastics pieces which are partially catalogued and digitised. The cataloguing was materials-based. Since plastics started being collected later than ceramics, which had already been assigned the letter P (for 'Pottery'), plastics were assigned the letter G, tellingly, the same initial as glass, suggesting they were acquired as an extension of the glass collection. This denial of plastics as a materials group within collections is further discussed in Chap. 13. Plastics identified so far are of British and Danish origin. They are found amongst British and international items of a very wide range that cover design and craft primarily for domestic consumption.

During its historical circulation between 1951 and 1976, artefacts were initially grouped according to material and later in themed displays. Thematic displays organised the objects in categories such as industrial design, folk design, craft and art. The main plastics displays bore the title *Plastics in the Home*. As the heading suggests, plastics objects in the Collection were mostly tableware, with arrays of cups, saucers, jugs, bowls, vases, and trays, plus pepper mills, bowls and insulated cups. Besides tableware, there were artefacts destined for other uses, either entirely made of plastics or with plastics parts. For example, the *Industrial Design and Engineering* display case included a set of plastics weighing scales (Fig. 12.1).

In the *Design for Play* case, there was the *Lock-a-Blocks* set of polystyrene bricks. In other groupings, we find a polyethylene waste-bin and samples of laminate for covering surfaces. Finally, at the more eclectic end of the Collection, there is a handful of decorative objects that may be seen to straddle the divide between design and art object, such as the 'trichromatic print on vacuum-formed polyvinyl chloride (PVC) with reflective metalised Melinex' (Museum Circulation Scheme, Metropolitan Archives London. I.L.E.A./S/LR/07/017).

In accordance with the instigators of the scheme, who aimed at promoting British manufacture as well as advancing the taste of 'the consumers of tomorrow' (Woodham 1996, 16), the displays presented a survey of plastics materials which exposed students to a selection of the synthetics mushrooming in the post-war period. Plastics in particular were seen as materials in need of framing within specific aesthetic parameters to compensate for the deluge of ill-conceived products. Based on the information derived from photographic documents, the range of plastics materials included objects made of 'vacuum-formed PVC; polythene; polystyrene; thermo-setting melamine-powder in a closed mould; Perspex; melamine;

Fig. 12.1 Travelling display case *Industrial Design and Engineering*, from the Camberwell ILEA Collection. Photo: Camberwell ILEA Collection

and acrylic' (Museum Circulation Scheme, Metropolitan Archives London. I.L.E.A./S/LR/07/017).

This brief survey of plastics in the Collection demonstrates their value as historical evidence at a moment of a fragile coalescence between design and manufacture. However, as it has been stressed elsewhere in this publication, 'value' is a broad term which can lend itself to multifaceted interpretations. In its current status as part of a University, the Collection brings pedagogical value to the learning of diverse audiences, including students and researchers. Apart from being rich in information that illuminates the state of design from the 1950s up to the 1970s, the Collection has also been used to enhance learning through the affordances of handling material culture and as the basis for a range of research, archival and curatorial projects (Fig. 12.2).

While the most prevalent linguistic connotation of value is monetary, I argue that recourse to funding and monetary expenditure do little to

Fig. 12.2 Plastics in the Camberwell ILEA Collection (clockwise from left): Insulex tumbler, Melmex lidded bowl, EKCO Nova cups and saucers, Opto Perspex artwork, Birchware small cup, Lock-a-block building blocks. Photo credit: Julia Parks and Ben Mullins

explain 'value' with regard to the group of plastics found in the Collection. As is the case with many educational services, and especially those of the past century, such collections occupy a conception of value that is mediated and mitigated by pedagogical concerns: in contradistinction to museum collections, artefacts were made freely available. This point is further supported by the administrators' relaxed attitude to damages: breaks and losses were approached as an inevitable consequence of the objects being handled. Though damages were recorded and where possible, items replaced by ILEA, there is no evidence that the objects were insured and no compensation was sought (Adams, Gene. Typescript of report on the Scheme for Design Appreciation. 1974. Camberwell ILEA Archive, University of the Arts London, UK).

Instead, documentation suggests that educational and contextualising texts exalted value beyond money: predominantly, the idea of value was bound up with aesthetics and taste so that the Collection's pedagogical value was filtered not through cost but through alignment to a specific, highly regarded, aesthetic canon. The foremost and original purpose of the Collection was to teach children the principles of 'good design' through the study of objects and materials (Foott 1952, 24–29).

Yet, it is important to keep in mind that discourse contained within the protected space of educational circulation, where objects could exist in a value bubble, only tells half the story. Outside the field of acquisitions for educational exhibits, the discussion will address how these same designs fared in what was becoming an increasingly crowded and competitive market place. Using the specific example of the *EKCO Nova* range, I will show how in the sphere of retail consumption, where production costs and prices matter, there emerges a different narrative for these items, one that ultimately undermines the rhetoric of modernism.

DESIGN COLLECTIONS AND THE 'GOOD DESIGN' MOVEMENT

In order to better understand how aesthetic value with regard to the Collection functioned, it is significant to place it within the context of the 'the good design' movement and those organisations that advocated its dissemination.

Due to the country's early industrialisation, there developed, in the United Kingdom, a succession of government-backed schemes aimed at improving the country's design output. For example, the Schools of Design, institutes which taught draughtsmen how to design for industry, were established as early as the 1830s. The need for such instruction rose partly out of state anxiety surrounding fierce international competition and the fear that through their own rapid industrialisation, neighbouring countries, particularly France, Bavaria and Prussia were catching up with Britain.

The belief that Britain was lagging in 'art appreciation', which in turn affected the quality of industrial output, was identified as a weakness and formed the basis of lengthy parliamentary debates (Hansard 1833).

This, coupled with utilitarian sensibilities about how the state could act for the benefit of the country, paved the way for interventionist taste agendas which shaped design education throughout the nineteenth and twentieth centuries. In terms of what 'tasteful' industrial output should look like, the early belief that public taste could, and indeed should, be improved by exposure to well-designed objects dovetailed with much proto-modernist rhetoric (Pevsner 1936).

A number of schemes implemented during the nineteenth and early twentieth centuries can be viewed as precursors of the Camberwell ILEA Collection. A few years after The Great Exhibition of 1851, objects made in the Central School of Design, the main School of Design in London, alongside examples from the newly established South Kensington Museums, were grouped in travelling collections. These were lent in rotation to provincial schools and provincial museums, marking the beginning of the Victoria and Albert Museum's Circulation Department, a project that inspired and eventually ran parallel to the schemes that supplied the Collection under review.

By the time of the Festival of Britain, the centenary celebration of the Great Exhibition, modernism had prevailed as the aesthetic that arbiters of taste valued above all else and the term 'good design' was coined to express it. 'Good design' had been a phrase used in the inter-war period to denote the aesthetic imperatives industrial production should follow (Board of Trade 1932).

To clarify the connotations of this term, 'good design' meant functionalism; that is, design fit for purpose and a restrained use of decorative elements. Though we should be mindful of oversimplifying, during the 1950s and 1960s the term 'good design' was used as a catch-all term, to imply a type of knowledge which was not rendered explicit, but rather determined by participation in the bona fide legitimate and legitimising institutions that advocated it.

Out of these institutions, the Council of Industrial Design (henceforth 'CoID') was arguably the most dominant and relevant to our discussion. The CoID, now the Design Council, is a government-funded institution, which was established in 1944 to support Britain's economic recovery after World War II. The CoID was directly involved with the Festival of Britain, which was the perfect showcase for championing British manufacture, new technologies and new materials. For years before the event, it had been selecting and assembling objects. The research yielded a list of 20,000 exhibition-worthy products and about half of those made it to the Festival (Crowther 2012, 8).

Plastics drew much interest during the Festival. Ernest Race's *Springbok* chair is a case in point. The chair had been designed specifically for the Festival, made of a metal frame covered with PVC for comfort and flexibility. The combination of metal and PVC proved how, in their mid-century incarnations, plastics were being coupled with more established materials to serve fitness for purpose. Such domestic designs, seen in the Festival but later also available on the market, were ideal vehicles where industry and aesthetics combined to stamp taste on millions of visitors, including the developing minds of visiting children.

In the aftermath of the Festival's success, the CoID sought to retain influence on how 'good design' was disseminated to the public. This belief led to the partnership with the London County Council in order to found the *Experiment in Design Appreciation*, which evolved into the *Circulating Design Scheme* and is now the Collection under discussion, which aimed to educate future consumers and designers. Indeed, as early as 1955, the CoID was campaigning for design education in secondary schools: 'There is growing desire on the part of education authorities and teachers for more information and material, which can be used in the teaching of design appreciation. Although this subject is not included in the General Certificate of Education, each year more schools recognise that design affects everyone's life and they come to the Council for information and guidance' (Council of Industrial Design 1955).

Evidently plastics fitted this agenda. It could be argued that they were the ideal materials to embody technological advancement and to intrigue a wide audience which would include future designers and engineers as well as consumers. Plastics' inclusion in the scheme reflected the materials' rapid expansion: while in 1929, only 8000 tons of plastics materials were produced in Britain, that figure had reached 110,000 tons by 1948 and it continued to grow exponentially (Farr 1955, 123). In the context of the Collection and beyond it, there was palpable optimism surrounding plastics. In conjunction with the ingenuity and entrepreneurship of British industry, plastics would not only herald a better, more functional world of material goods but would also benefit the economy by creating new career paths for ambitious youths (Farr 1955, 125–129).

Thus, plastics started being collected early on: the earliest documentation in the Collection is found in the 1950s *Materials* display. As the text panels document, there were samples of 'aminos, phenolics, acrylics, cellulosics', linked to depictions of consumer or industrial applications. Plastics also appeared in the *Packaging* display, where plastics packaging solutions were showcased alongside paper-based packaging (Museum Circulation Scheme Metropolitan Archives London. I.L.E.A./S/ LR/07/017).

Despite recognising the role of production, increasingly the Collection's preoccupation was to showcase the best of what was available on the market. The London County Council was funded by government grants and it was run by bureaucrats and education managers with varying degrees of personal interest, knowledge and involvement in design and aesthetic pursuits. With regard to the Experiment in Design Appreciation, the Collection's administrators cherished their close alliance with the CoID, as the authority to trust in matters of selecting 'the best'. Though the CoID had participated in committees where the initial groups of objects were assembled, it ceased its involvement in 1956, possibly because in the same year the CoID opened its own showroom in London.

Therefore, from 1956 onwards, the London County Council was the sole administrative body managing the Collection. That is not to say that the CoID's influence on selection diminished. Throughout the history of collecting, the administrators of the Collection continued to depend on the CoID's aesthetic choices. These choices were made explicit through the three main vehicles that the CoID implemented to disseminate its message: its own publication, *Design* Magazine (1949–1999); its showroom in central London (1956–1989) and especially, the yearly *Designs of the Year* awards (1957–1988).

DESIGNS OF THE YEAR IN THE CAMBERWELL ILEA COLLECTION

The *Designs of the Year* scheme had been initiated in 1957 'in order to improve design standards [...] to promote the best British designs [...] and to encourage retailers to stock [them]' (Crowther 2012, 7). *Designs of the Year* were derived from the *Design Index*, the list of British products in which the CoID invited manufacturers to participate. In addition to a committee choosing products, the CoID enjoyed the patronage of the Duke of Edinburgh, who awarded his own separate prizes for 'elegant design'.

The awards' emphasis on domestic products encouraged their acquisition by travelling collections: alongside the London County Council, the V&A and smaller provincial schemes were also acquiring awarded designs. It helped that the focus of awards was overwhelmingly on smaller products, though later on the Council and the Duke expanded the awards' scope to include such projects as equipment for manufacture and sailing vessels. While British industry and design were booming, CoID accolades were bestowed on products drawn from a limited pool. Inevitably, the Council attracted criticism that familiar names appeared to be winning year after year: designers like David Mellor and Robin Day or manufacturers such as Hille, Wilkinson Sword and Wedgwood were among those that had won multiple awards (Stewart 1987, 191).

This careful aesthetic policing is also reflected in the Collection: items were sourced from a limited number of purveyors. It is therefore not surprising that many of the plastics in the Collection, including the *Melmex* and *EKCO Nova* ranges of tableware as well as the Collection's numerous *Insulex* tumblers, can be identified as prominent *Designs of the Year* winners: *Melmex* in 1957, *Insulex* in 1964 and *EKCO Nova* in 1968. Other plastics selected for inclusion in the scheme, such as the Birchware range (despite the natural connotations of the name, these were injection-moulded plastics produced by the English company Birchware Ltd) and the Danish items produced by Torben Ørskov & Co, were likely to have been presented in CoID-supported platforms such as the CoID's showroom in London's Haymarket or through articles in *Design* magazine. In addition to these CoID-controlled spaces, the Collection's administrators also acquired items from a handful of upmarket London retailers, particularly Liberty's, Heal's and Primavera. The purchasing strategy for plastics circumscribes a rather contained taste field which appears to be at odds with the era's trend for technological and retail expansion.

THREATS TO THE MODERNIST CANON

Despite the administrator's effort to ring-fence acquisitions within the modernist aesthetic, this became an increasingly difficult task as time went on. It is important to keep in mind that the collecting and circulating of plastics for the Camberwell ILEA Collection spanned nearly three turbulent decades (from 1951 to 1976). The Collection's transformation over the years reflect varying cultural values; its trajectory traces changes in British society at large. During the time in question, dominant attitudes to British design underwent a notable taste shift from post-austerity and functionalism to the advent of pop and its engendering culture. Researching beyond the 'good design' camp helps disclose the complexities within the story of post-war British design, particularly pertinent to plastics in the Collection.

I have shown how at the Collection's inception, the project of 'good design' emerged as a government-funded imperative: an imperative where modernism was pitted against non-canonical design and the fickle taste of the consumer. By the mid-1960s, this mandate had turned into a race against time: sources from the time hint at an urgency to convince manufacturers and consumers of the benefits of 'good design' before modernism collapsed under the weight of its own untenable expectations (Meikle 1997).

It is telling that by the early 1970s, the Design Council decided to remove 2000 products from their Design Index, the list of about 9000 Council-approved products of good design, in order to 'allay two suspicions of the public: firstly that good design is expensive and secondly that aesthetics and efficiency are somehow mutually exclusive' (Stewart 1987, 224). The indication that the public viewed, by the turn of the decade, 'good design' with suspicion, as representing a paternalism that was being rejected and questioned with regard to its aesthetics and functionality, implies also a rejection of what the whole project of modernism had stood for.

From the mid-1950s to the mid-1960s, as the contingency of plastics within the Collection grew to hundreds of objects, the strict taste filter led to products that had won the CoID's approval. This allayed worries of diluting the Collection's purpose and message. Yet, outside the confines of the Collection, and while the plastics industry continued to develop, differing voices had gained momentum. As Jeffrey Meikle has pointed out, 'plastics, by its very nature, complicates efforts to think about it' (Meikle 1997, 279). Having managed to overcome the bad publicity associated with the shortcomings of plastics products hastily introduced to the market right after the end of World War II, plastics steadily gained the consumers' trust during the 1950s. However, by the 1960s, plastics had attracted the criticism of being 'ersatz' materials (Katz 1984, 13) and in the 1970s it would be fair to argue that appreciation of plastics followed an inverse trajectory to its expansion. While people used more plastics each year, its reputation steadily declined. Schemes like the Collection strove to counteract the fluctuation in consumer loyalty; however, the navigation of aesthetic agendas was becoming increasingly treacherous.

EKCO NOVA: THE BURDEN OF 'GOOD DESIGN'

Dissidence in the mid-1960s went beyond consumers' perennial ambivalence towards plastics: the proliferation of voices against authority in every part of society meant that aesthetic approval from government-run institutions was becoming problematic. Moreover, that approval was not enough to make plastics products commercially viable. By the 1960s, it had become evident to British manufacturers that in order for plastics to yield profit and recuperate the high initial investments in research and development, large volumes of sales were necessary (Farr 1955, 125). While early polymers were simpler thermosets and semi-synthetic, the advent of composites in the post-war years caused a bewildering explosion of the range of polymers and their capabilities. Industries anticipated huge growth and there was generous investment in the development of patentable polymers.

Faced with a multitude of alternatives, producers needed to make the right choice for their purposes. One such novel composite polymer was Styrene-acrylonitrile or SAN, out of which the celebrated range of British tableware *ECKO Nova* was produced. The parent company, EKCO, had set its ambitions high for this product: at the time, the plastics of choice for domestic items, and especially kitchenware, was melamine.

In February 1968, ILEA purchased about forty *EKCO Nova* items. Evidence in the documentation proves that these were purchased from Primavera, the upmarket supplier on London's Sloane Street (Scheme for Design Appreciation record card, 1968. Camberwell ILEA Archive, University of the Arts London, UK). In view of understanding how 'good design' was validated through socio-cultural contexts, it is telling of *EKCO Nova*'s distinguished status that a gallery such as Primavera, best known for its studio pottery exhibitions, had decided to stock the range. The set was also carried by Habitat and Heal's, two more retailers trading at the high end of the interiors market.

However, *EKCO Nova*'s is a cautionary tale of how aesthetic values may undermine a product's economic sustainability; it is also a tale that challenges modernist beliefs, among others the belief that 'good design' eventually pays off its initial investment (Fig. 12.3).

EKCO Nova's designer was David Harman Powell, who had risen to be one of the country's best-known industrial designers and much admired amongst his peers. Harman Powell had worked for E. K. Cole plastics in his twenties, then moved to British Industrial Plastics (BIP) before returning to E. K. Cole as chief industrial designer in 1960, by which time the company was renamed EKCO. His association with EKCO lasted into the early 1970s. Harman Powell's credentials made his work sought after in industry as well as in the higher education sector: London's Royal College of Art appointed him its first tutor in plastics.

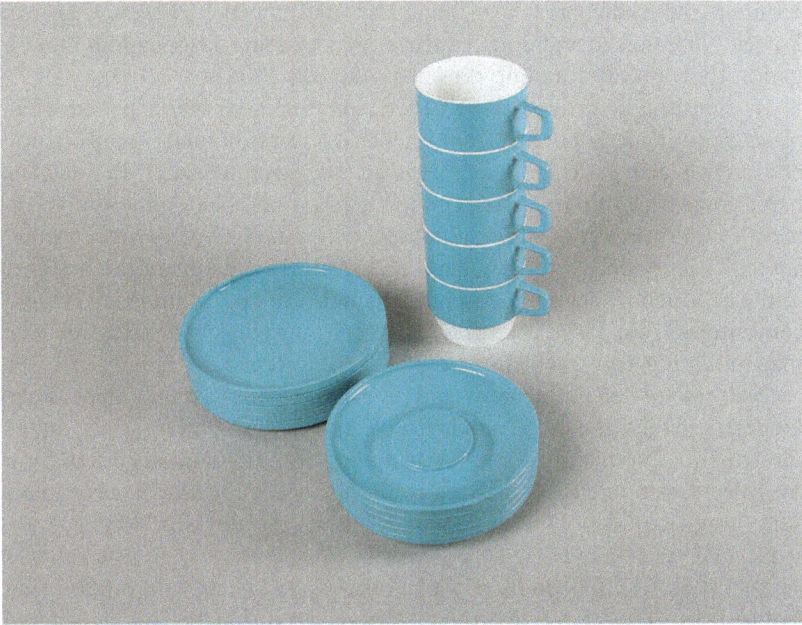

Fig. 12.3 *EKCO Nova* tableware designed by David Harman Powell for EKCO, about 1968. Photo: Museum of Design in Plastics, Arts University Bournemouth

As a member of the BIP's design service, Harman Powell had already been involved with re-imagining the *Melmex* range of tableware, which had previously been produced for Midwinter in the late 1950s. At the time, BIP, having acquired knowledge from its U.S. counterparts, held a near monopoly on melamine, through supplying industry with Melmex, a patented type of the material. Despite their continuing success as a ceramics manufacturer, Midwinter had been keen not to miss out on the plastics bandwagon. It is important to keep in mind that in terms of tableware, the 1950s can be identified as a time when plastics presented as serious contenders to ceramics. This gave rise to some anxiety on the part of porcelain, glass and ceramics manufacturers, even the most established ones.

Originally conceived by BIP's Albert Woodfull and John Vale, Midwinter *Melmex* made the 1957 selection of CoID's *Designs of the Year*. Naturally, *Melmex* tableware was acquired by both the Victoria and Albert Museum and the Circulating Design Scheme. When Ranton and Co.

acquired the trademark for a similar material named Melaware, Harman Powell got involved and updated the design by using a two-colour combination through bi-moulding the cups (Akhurst 2004).

By the time he arrived at EKCO, Harman Powell was experienced in plastics production. He held the belief that success hinged on production efficiency, placing particular emphasis on the planning stages. Besides, the nature of production in plastics rendered inevitable the designer's involvement in every aspect of the product cycle. The *EKCO Nova* project guaranteed synergy between design and technology by involving the designer in the production planning committee and on the company management committee. John Heyes, writing in *Design* magazine in 1967, explained that moulding tools, which were made of hardened steel, needed to be precise, since later refinements and alterations were virtually impossible. Understandably, the steel moulds were the largest outlay of the initial investment. The margin for error at design development stage was zero: the injection-moulding technique could not be reproduced in prototype. The design could not be proven right until 'the die was cast, in every sense' (Heyes 1967, 47).

In 1965, EKCO conducted a market study asking the public what would be an ideal range for them. It transpired that the public wanted 'space-saving, stable, easy to clean and handle, nestable, stackable, modern-looking, functional and comparatively inexpensive' wares (Council of Industrial Design 1968, 27). Such feedback justified the choice of SAN as the plastics to be used in the *Nova* range, instead of widely applied melamine. The decision was based on the fact that this costly new material displayed a range of qualities that improved on melamine's performance: the end product would be rigid and resistant to fracture and in stark contrast to melamine, it would be resistant to staining.

In addition, the design anticipated the common problem of surface chafing which characterised stacking plastics of the time. The main bearing surfaces of the *EKCO Nova* range were transferred to the rims and edges of the items. Diameters of products were standardised into two groups with the saucer acting as an inter-stacking link. Production of *Nova* could be speeded up by injection moulding since SAN was a thermoplastic rather than a thermosetting material. Injection moulding would be cost-effective if production runs were sustained long enough to recoup the initial outlay of steel moulds.

With regard to colour schemes, transparent tints as well as opaque colours could be achieved in SAN, making it easier to produce a range based on various colours but using the same moulds. Furthermore, moulds were designed to be as adaptable as possible: the mould used for the cup was the same as the one used for the milk jug. The body forms of the two items were identical, while the differentiation of the cup was achieved by the attachment of a handle requiring a small separate mould.

As a sign of its belief in the project, EKCO produced leaflets and advertisements, which emphasised the link between the high quality of design and the longevity of the products. The range came with a guarantee for one year from the date of purchase which covered the problems that had previously plagued the image of plastics tableware: according to the manufacturers, cracking, chipping and breaking did not affect *EKCO Nova* tableware.

Indeed, in terms of technical and design accomplishment, this was a ground-breaking series of products. Upon awarding the Duke of Edinburgh's *Prize for Elegant Design* in 1968, the judging panel commented on *EKCO Nova*'s avoidance of the established tendency to use forms originally developed from ceramic materials, a tendency that the panel frowned upon as 'ill-adapted to newer materials'. Instead, they applauded *Nova* for its forms, which were 'both pleasing and entirely suitable to the material and the colours in which it can be made.' They concluded that *EKCO Nova* had achieved the practical qualities of convenience and durability with shapes and colours of 'precision and elegance' (Council of Industrial Design 1968, 27). EKCO utilised the accolades bestowed on its *EKCO Nova* ranges in their marketing and advertising: they placed numerous advertisements in *Design* magazine, mentioning the Duke of Edinburgh's prize.

However, the praise that *EKCO Nova* attracted from industry and the various 'good design' platforms, including its acquisition by the Collection, follow a value narrative freed from the consequences of economics. Despite its image as a highly successful company, EKCO actually maintained a rather precarious financial balance led by a seasonal production cycle much affected by the slower summer months. Its fragile position in the market is attested by its turbulent history during the 1960s and 1970s, including mergers with PYE (1960) and the European Phillips Electrical in 1967.

Furthermore, both the *EKCO Nova* tableware and the subsequent cutlery range have been described as 'financial disasters' as they failed to realise the sales volume EKCO needed to make them profitable. David Oxley, polymer technologist and former employee of EKCO plastics, explained that the manufacturers had relied heavily on their partnership with British Airways (BA) who would use the range for their First Class service. Crucially, Oxley maintained that *EKCO Nova* did not fulfil BA's expectations of prestigious tableware. This comment is indicative of the difficulty plastics encountered in infiltrating dominant taste: 'The real reason that the Nova range was not successful was that the BA didn't like them as they were not "posh enough" for first class and too expensive for economy class passengers. In those days First Class passengers used metal cutlery and the SAN cups soon scratched with metal spoons and the plates showed cut marks with metal knives. Once we lost the BA contract (they only took the first batch if I remember rightly) the volume wasn't enough to cover the costs and our domestic housewares division at EKCO couldn't get the sales up' (Personal communication with author, 2013).

Oxley's comments and the case study of *EKCO Nova* and its highly commended products expose the historiographical problem at the heart of the rhetoric of 'good design': the bias towards 'well-designed' products could not be relied upon as predictors of commercial success; that was despite the continuing effort to 'educate' consumers and cultivate a perception of certain plastics as modern equivalents to porcelain and stainless steel.

Looking closely at *EKCO Nova*'s trajectory in the marketplace provides an illuminating counter-narrative of how such commodities fared beyond the Design Centre showroom. In this context, it makes sense to refer to the cost: SAN was a superior but ultimately expensive material, and it was applied in quantities only explained by the designer's wish to have thicker parts and render the end product more pleasing to the eye and to handle (David Oxley, personal communication with author, 2013). According to a 1968 price list in the company's archives now part of Harman Powell's papers at the V&A Archive of Art and Design, *EKCO Nova* actually cost more than most of its contemporary china at the time (EKCO Nova price list. 1968. ADD/1994/8/41/1-18. Victoria and Albert Museum Archive of Art and Design, London, UK). EKCO tried to justify the high prices by enlisting its design credentials. In its marketing, the argument for superiority over rivals hinges on the aesthetic praise bestowed upon the company's products. '*EKCO* is not cheap. We know. But it lasts and lasts. And lasts. However bad you treat it.' (Promotional Leaflet. 1970. ADD/1994/8/41/1-18. Victoria and Albert Museum Archive of Art and Design, London, UK.)

Observing from the vantage point afforded to the design historian, we are aware of the cultural paradigm shift that the company was not alert to:

while EKCO was busy making durable tableware, plastics were emancipating from a status where they had acted as replacements for other materials to establishing their own market conditions. Moreover, with the advent of 'popular' or 'pop' design, durability was taking second place to expendability. Despite its inherent and invested design value, *EKCO Nova* was a commercial failure and a case when the rhetoric of 'good design' fell short of convincing consumers.

CONCLUSION

This examination of case studies from the Camberwell ILEA Collection points to a crucial moment in the history of domestic plastics. As educational objects, they aimed to serve two explicit purposes: firstly, to inculcate a certain aesthetic and a taste for 'good design' that would shape domestic consumption alongside the principles of the functionalist modern style. Secondly, to excite students about the potential of working as engineers and industrial designers, through contact with the canonised 'best' of British and international manufacture.

However, my research suggests that in many cases, and despite concerted efforts, neither the authorities of design education, nor the manufacturers were able to carve for these products the niche that they had desired for them. It was plastics materials themselves, their very versatile nature, which ultimately dictated their cultural value and their place in history.

The *EKCO Nova* range demonstrates that it proved ill-advised to treat the design and marketing of domestic plastics products as prestigious, timeless objects that justified high price tags. The failures of the plastics industry in the tableware sector renewed the relevance of porcelain and further validated it as the aspirational material of consumer choice.

That is not to say that there are no successful examples of costly, coveted plastics products for domestic use dating from the late 1960s. However, some of the most enduringly celebrated design in plastics did not acquiesce in the modernist canon but challenged and subverted it. In the adventure playground of the nascent postmodern design movement, plastics came into its own. Many successful designs in plastics were achieved by those who had embraced the material on its own terms and had exploited its idiosyncrasies, for example, Vernon Panton's S-shaped chair made in a multitude of colours.

Over time, the majority of materials we categorise as plastics eventually arrived at a position of expendable, indispensable, intrusive ubiquity, indeed the semantic opposite of the exclusivity to which *EKCO Nova* and other examples from the Collection had aspired. This happened swiftly and irrevocably. Once the value of plastics as cheap, accessible materials was embedded in the collective psyche, it was increasingly hard to defend elite notions of 'good design'. Plastics in the Camberwell ILEA Collection capture a moment in design history when 'good design' was in need of a re-framing that would liberate it from modernism's aesthetic constraints. Plastics offered an opportunity for such re-framing. After all, plastics provide affordable, yet often superior, solutions for a bewildering range of uses and they have popularised design in ways that may not always fulfil the aesthetics of modernism but which have ultimately delivered modernism's democratising mandate.

NOTE

1. In 1976, ILEA's *Circulating Design Scheme* had its funding drastically cut and the objects had been in storage until they were bequeathed to Camberwell College of Arts in 1990. It was a way of preserving the *Collection* and saving it from being dismantled as *ILEA* was disbanded.

BIBLIOGRAPHY

Akhurst, Steve. 2004. The Rise and Fall of Melamine Tableware. *Plastiquarian* 32: 8–10.

Board of Trade. 1932. *Report for the Committee on the Production and Exhibition of Articles of Good Design and Everyday Use*. London: HMSO.

Crowther, Lily. 2012. *Award Winning British Design*. London: V&A Publishing.

Council of Industrial Design. 1955. *Tenth Annual Report for the Year 1954–1955*. London: HMSO.

———. 1968. Duke of Edinburgh's Prize for Elegant Design. *Design* 233: 27–28.

Farr, Michael. 1955. *Design in British Industry: A Mid-Century Survey*. London: Cambridge University Press.

Foott, Sydney. 1952. Design Appreciation in London Schools. *Design* 49: 24–29.

Hansard Parliamentary Debates. 1833. HC Deb. 20, 139–74. July 30. https://api.parliament.uk/historic-hansard/commons/1833/jul/30/national-education#S3V0020P0_18330730_HOC_21. Accessed 19 May 2020.

Heyes, J. 1967. Getting It Right the First Time. *Design* 217: 46–53.

Katz, Sylvia. 1984. *Classic Plastics*. London: Thames and Hudson.

Meikle, Jeffrey L. 1997. Material Doubts: The Consequences of Plastic. *Environmental History* 2 (3): 278–300.
Pevsner, Nikolaus. 1936. Reprint, 1991. *Pioneers of Modern Design*. London: Penguin.
Sparke, Penny, ed. 1990. *The Plastic Age: From Modernity to Post-Modernity*. London: Victoria and Albert Museum.
Stewart, Richard. 1987. *Design and British Industry*. London: John Murray.
Woodham, Jonathan. 1996. The Consumers of the Future. In *The Camberwell Collection: Object Lesson*, ed. Jane Pavitt, 16–17. London: The London Institute.

CHAPTER 13

The Changing Fortunes of Plastics in Museums and Galleries

Deborah Cane and Brenda Keneghan

This chapter is based on the personal experiences of two museum professionals: a conservation manager working at Tate[1] who formerly worked at the Birmingham Museum Trust (BMT) and a scientist working at the Victoria and Albert Museum (V&A). It explores a path to acceptance of objects made of plastics or with plastics parts in museums and art galleries; how they may be differently valued by different museum professionals; and how value perceptions of plastics things may impact on the way such artworks and objects are treated (or mistreated) within the museum context.

D. Cane (✉)
Tate, London, UK
e-mail: Deborah.cane@tate.org.uk

B. Keneghan
Victoria and Albert Museum, London, UK
e-mail: brendakeneghan@icloud.com

© The Author(s), under exclusive license to Springer Nature Switzerland AG 2020
S. Lambert (ed.), *Provocative Plastics*,
https://doi.org/10.1007/978-3-030-55882-6_13

235

'PLASTICS DENIAL SYNDROME'

The term 'museum' is derived from the Greek 'Mouseion': a temple, seat or shrine of the Muses. Traditionally, museums have been places where things that are old, rare and valuable are collected. These three adjectives used to describe the typical museum object are seldom used to describe things made of plastics. Likewise, galleries have traditionally been places where art that is highly valued and therefore usually expensive is collected and exhibited. Thus, the wide understanding of both museums and galleries is at odds with the frequently held view of plastics as ubiquitous, cheap and disposable materials that plunder the world's resources.

When Brenda Keneghan joined the V&A as a polymer scientist some twenty-five years ago, she was asked to undertake a survey of all the collections to see which plastics were lurking amongst them and what condition they were in. This mammoth undertaking had been rumbling on for a long time, initiated in response to the sudden degradation some years previously of a Naum Gabo sculpture in the Philadelphia Museum of Art and the model for it held in Tate's collections. The sculpture in question *Construction in Space: Two Cones* was made from the unstable plastic, cellulose acetate (Mundy 2020). Museum professionals were beginning to realise that in spite of popular opinion, plastics do not last forever and the failure of the sculpture was a very costly example of this phenomenon.

Most of the plastics objects that Keneghan located at that time were rather mundane, not at all as glamorous as costly sculptures. They were mainly toys from the Museum of Childhood at Bethnal Green and early semi-synthetic imitations of natural materials in the form of jewellery or handbags. The survey also covered decorative parts of costume in the Theatre Museum collection, many objects in the Furniture collection and cups and decorative containers in the Ceramics collection. With very few exceptions, the impression given by curators was that the objects were of low value and low importance; certainly, plastics as a materials group was looked down on at that time by many people in the V&A. When she approached keepers of collections, as heads of departments were called, to gain access for her survey, she was often met with what she termed in an article entitled 'Plastics—Not in My Collection', published in the *V&A Conservation Journal*, the 'plastics denial syndrome' (Keneghan 1996). Several keepers were vehement in their responses, denying the existence of any plastics in their collections, as if she were accusing them of having inferior objects.

Around the same time another, then, V&A staff member, Susan Lambert (editor of this book), was also having an issue with attitudes to plastics. This concerned the acquisition of objects for the V&A's first global *20th Century Gallery*. Finding that certain sorts of mainly electrical products that tended to be encased in plastics had not been collected, she was given a budget to address this gap. A particularly challenging acquisition proved to be a newly designed vacuum cleaner made out of bright pink and purple plastics: the *Dyson G-Force Cyclonic* vacuum cleaner. The designer James Dyson was determined that his latest invention had the look of Nasa Technology but, unable to find a British manufacturer, he went to Japan. Thus, its unusual colours were dictated by Japanese taste. The use of plastics, their colour, texture, transparency, and weight all contribute to the iconicity of this object but not everyone agreed. The vacuum cleaner was to form part of the Metalwork collection as, by tradition, vacuum cleaners were metal objects; however, according to Lambert, 'the Metalwork Collection kicked and screamed' (conversation between Keneghan and Lambert at the time) as they did not perceive such an object as appropriate for their collection. Subsequently, perception has changed. The vacuum cleaner had pride of place mounted on a pedestal adjacent to a glistening e-type Jaguar in the V&A exhibition, *British Design from 1948: Innovation in the Modern Age*, 2012 (Breward and Wood 2012, 276–277; V&A n.d.-a).

The lack of respect for plastics, in the then context, is nevertheless understandable. The Metalwork Department's primary focus was on precious metals, valuable even before they become objects of desire, whereas most plastics as such have no inherent value. Even the pewter collection, of considerably less financial value than the gold and silver collections, but when melted down worth much more than plastics, was treated to some extent as a poor relation: it was curated by a less senior staff member and the gallery was the last of the Department's materials galleries, in a cycle of renewal, to be refurbished.

This lack of respect was also a result of timing. Decorative arts museums, like the V&A, burgeoned all over Europe in the second half of the nineteenth century (Burton 1999, 112–113). Their collections tended to be organised according to a system promulgated by the architect and architectural historian, Gottfried Semper. Semper argued that form was determined by materials and therefore the ideal organisation for such collections was not by chronology or aesthetic significance, as was usual at the time, but by the different arts as dictated by the four primordial techniques of making: the weaving, woodworking, ceramic and metal-working arts (Hvattum and Hermansen 2004, 128–129). Remarkably, in this respect, the V&A was reorganising its collections according to this system in 1909, the very same year that Leo Baekeland patented his discovery of the first fully synthetic plastic to be invented, phenol formaldehyde trademarked Bakelite. It was to take several decades before plastics became a significant materials group. Thus, it is understandable that the reorganisation did not include a plastics department. It is however noteworthy that that remains the case today. Instead, plastics are dispersed around the museum according to the material type of which the object was made before the advent of plastics, and while, for example, swathes of the V&A's top floor are devoted to explaining the manufacture of ceramics, there are no permanent galleries dealing with manufacture in plastics.

Bearing in mind that authenticity is at the heart of museum and gallery collections, the dismissive attitude may have stemmed also from the view that plastics were considered to provide mainly cheap imitations (as they often did) of more precious natural materials. This aspect of plastics production is discussed at greater length elsewhere (Chaps. 1, 6, and 14) in this volume. An exacerbating factor was that after World War II, when plastics began being used for their own sakes, the fledgling plastics industry was somewhat cavalier, not always making quality its highest priority. Sometimes the products did not last. For example, early polyethylene washing-up bowls suffered from a fundamental flaw: environmental stress-cracking. This meant that in a hostile environment (in this case hot soapy water), the residual stress in the material was accentuated and eventually the material failed (Fig. 13.1).

Unfortunately, many plastics objects of the time were not fit for purpose making the 'plastics denial syndrome' in a museum focused on good design understandable.

Plastics Products as Collectable Museum Objects

However, there were also furniture makers who seized upon the unique properties of the various plastics as materials in themselves rather than as imitators. Some of the earliest pioneers of modern furniture saw their potential as, for example, Charles and Ray Eames in their early moulded plastics armchair manufactured by Zenith Plastics from fibre-reinforced polyester and distributed by Herman Miller (V&A n.d.-b). Zenith Plastics had made fibreglass casings for radar machines during World War II and the Eames had made many moulded plywood aircraft parts and

Fig. 13.1 Polyethylene washing up bowl showing stress cracking, 1949. (Photo: Plastics Historical Society and Museum of Design in Plastics, Arts University Bournemouth)

subsequently one-piece moulded chairs from plywood. The plastics chair was, therefore, an obvious next step that demonstrated how wartime military technology found peacetime civilian applications.

In the same issue of the aforementioned *V&A Conservation Journal*, there is an article by Gareth Williams, then a curator in the V&A's Furniture and Woodwork Department and now Head of Design at Middlesex University. He presented a different attitude writing in admiration of the use of plastics in many different ways and the exciting furniture that they could make. He showed how the properties of plastics and polymer foams had been and were being exploited to produce furniture in forms that would be impossible in any other materials (Williams 1996).

A third article in the journal presented yet another view. Roger Griffith, then a RCA/V&A conservation student and now a conservator at the Museum of Modern Art, New York, described the problems faced by conservators with plastics-based pieces of furniture less than thirty years after their manufacture. Many of the objects discussed by Griffith were by iconic Italian designers who were amongst the first to explore the possibilities of synthetic materials in furniture, for example, the inflatable 1968 *Blow* chair designed by the Milan-based studio of De Pas, D'Urbino, Lomazzi and Scolari and made in transparent polyvinyl chloride (PVC), the first inflatable chair to be mass-produced (V&A n.d.-c). It was in poor condition although acquired soon after its original manufacture (Griffith 1996). The frequency of such issues relating to the plastics objects that did enter the V&A's collections may well have limited enthusiasm for collecting in this area (Fig. 13.2).

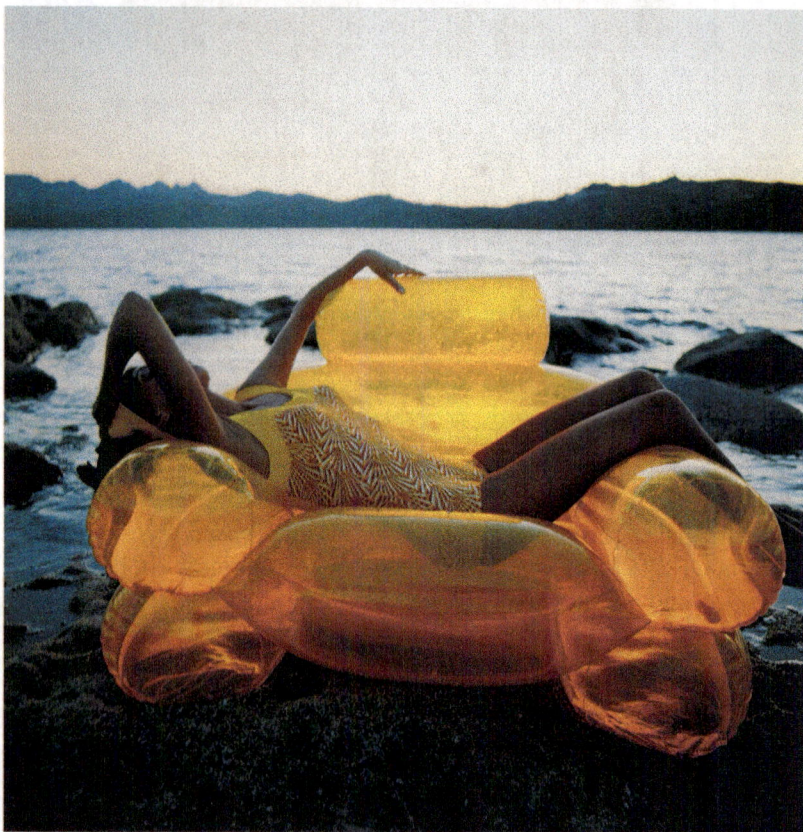

Fig. 13.2   *Blow*, designed by Jonathan de Pas, Donato D'Urbino, Paolo Lomazzi and Carla Scolari, manufactured by Zanottta SpA. (Photo: Courtesy of Zanotta spa)

However, given that the V&A claims to be 'the world's leading museum of ... design' and that for nearly half a century more, and a greater range of things, have been made of plastics than of any other materials group (Cascini and Rissone 2004), it is questionable whether even now it does justice to the diverse achievements of mass-produced design in plastics. Nonetheless, it is rich in outstanding pieces, especially chairs, which demonstrate plastics' potential fundamentally to rethink design, even if made only in small runs.

Notable among them is Gaetano Pesce's *Pratt chair* (1983–1984) (V&A n.d.-d) which more or less takes the form of a chair but is foremost an investigation of the material. It was produced for the Pratt Institute in New York as a series of nine experimental chairs made by hand-injecting increasing densities of urethane resin into a mould, each successive version using the same process of production, but captured in a temporal moment of development. Initially, not all the chairs were self-supporting and now some have collapsed while others have hardened to the point of uncomfortable resistance for the sitter. Pesce's idea that an object might offer an elucidation of its own eventual obsolescence or decay was key to the postmodern strategies of the period and, in spite of their fate, they are perceived as iconic. In addition to the one in the V&A there is one in the Museum of Modern Art, New York, (MoMA n.d.), three at the Vitra Design Museum, Weil an Rhein (Vitra Design Museum n.d.), and No. 7 in the series realised 15,000 euro at the Dorotheum auction house in 2016 (Dorotheum 2016).

By contrast, the *Materialised Sketch* chair (V&A n.d.-e) is an investigation of process and poses the question: is it possible to design directly onto space? Front Design has developed a method of creating freehand sketches in three-dimensional form by combining two advanced techniques. Pen strokes made in the air are recorded with Motion Capture and become 3D digital files; these are then 'materialised' through 3D printing, a method of manufacture discussed in greater depth in Chap. 6, into real pieces of furniture in an acrylonitrile butadiene styrene (ABS)-like resin strengthened with tiny particles of ceramic. Plastics are in this case the material capable of giving form to the idea (Fig. 13.3).

Use of recycled plastics in design is also represented in the V&A's collections. Foremost again are chairs (there is a different inquiry required about object types thought to be collectable by museums also mentioned in Chap. 10), for example, one designed by Jane Atfield and another by Bär + Knell. Jane Atfield's *RCP2* of 1992, known widely as the 'Atfield',

Fig. 13.3 *Materialised Sketch* by Front Design, 3D-printed by Alphaform in ABS-like Resin, 2005. (Photo: Front Design)

is made from high-density polyethylene bottles, previously used for anything from shampoo, milk, detergent and sun oil, which were collected in community recycling schemes, washed, chipped and pressed into various thicknesses using redundant plywood presses. The chair was sold by Made of Waste in quite considerable numbers: the largest order of thirty chairs was for Live TV at Canary Wharf (V&A n.d.-f). The Bär + Knell chair was commissioned by the V&A and made from waste packaging collected by Museum staff in June 1996. Although other versions of the chair exist, each is characterised by the nature of the recycling making each one unique (V&A n.d.-g).

PLASTICS IN CONTEMPORARY ART
IN THE GALLERY CONTEXT

Given the extent of plastics' role in design, it is not surprising that a museum of design such as the V&A should include objects made of plastics in its collection even if in a different way from its representation of other materials groups. However, plastics are less established or more hidden as 'gallery' materials. Nonetheless, as noted in Chap. 3, the extent of plastics' contribution to sculpture is considerable and is found extensively in a wide range of contemporary art. Derek Pullen, formerly Head of Sculpture Conservation at Tate estimated that in 2009 its collection contained some 300 sculptures with a high plastics content although had he included film and video works and all paintings, prints and photographs on polyester, or those using acrylic, PVA and alkyd paints the number would have been very much greater (Lambert 2012, 169).

Contemporary art has been described as 'a hybrid of sculpture and art, an amalgamation of materials, media and techniques. It borrows from cultural forms outside of art—"life"' (Ferriani and Pugliese 2013, 9). This quotation is a perfect description of *ARTicle 14, Débrouille-toi, toi-même!*,[2] an artwork by Romuald Hazoumè (b. 1962) acquired by BMT in 2012. Hazoumè was the 2007 winner of the prestigious Arnold Bodé Prize (Documenta 12, Kassel, Germany), which catapulted him into the top ranks of the international artistic community and made him unique amongst African contemporary artists (October Gallery n.d.). He says of his work: 'I send back to the West that which belongs to them, that is to say, the refuse of consumer society that invades us every day.' His work thus addresses key issues: 'western' domination, consumerism and our throwaway society (Donoghue 2019) (Fig. 13.4).

Physically, *ARTicle 14* depicts a street-trader's cart from Benin, which would normally carry a range of soft drinks, beer, plastics toys, brushes, footballs, razors, and pans. Hazoumè, playing with the concept of a standard street-trader's cart displays goods of a similar kind, but ones that have been gathered from things that have been thrown away in the West. They are objects made redundant by technological advance and the upgrading of new products, including an iconic Nokia Phone, Nike trainers, Apple laptops, Microsoft hard drives and a pair of faux leather trousers. These are articles that many museum visitors will have disposed of and that more 'developed' countries send to less 'developed' ones. The sculptural piece is not a real cart but rather made of a metal frame that is bolted together to

Fig. 13.4 *ARTicle 14, Débrouille-toi, toi-même!* by Romuald Hazoumè. On display at Birmingham Museums and Art Gallery. (Photo: Birmingham Museums Trust)

create the 'cart' on to which are secured wheels. Placed inside, hanging and fixed to the outside of the 'cart' are 711 objects held in place by wire, string, sticky tape and sometimes by the object itself, such as the shoelaces of a set of trainers. Of the 711 items, approximately 320 are primarily made of 'plastics' and nearly all of them, about 90%, have plastics components such as rubber hand grips on golf clubs, cellophane covers to books and sheaths over metal wires. The artwork also includes further elements that add context: two large photographs of a street market, and sound recordings and a video shown on a historic (1980's) portable 'plastics' television.

Attitudes towards *ARTicle 14* are varied and often informed by the familiarity of its content. Visitor comments range from: 'I don't get it... I had one of those [as they head towards the artwork to handle said object] ... did we pay good money for that ... it's fine but it's just a cart ... wow look at all those plastics' (Conversations with the Visitor Services Team, who were on duty in the gallery, and the Art Handling team). Museum staff responsible for its curation and preservation are inevitably influenced in their attitudes by the key objectives of their specialisms. Curators tend to prioitise its interpretation to the public and conservators the maximisation of its useful life. Additionally, both have to take into account the artist's intentions and may in this case be affected by his tendency to refer to his artwork as a cart (Jo Digger, Artist's Interview, New Art Gallery Walsall, 26 March 2013, Birmingham Museums Trust acquisition files), inclining them to treat it like a cart rather than as a work of art.

There are many factors associated with a work of contemporary art made of modern materials that may lead different types of museum professionals to make different decisions about its care. Protocols may be interpreted differently. Because the art work is 'new', some may assume it is in good condition. The fact that it is a 'cart' incorporating familiar low-cost items may disguise what is involved in preserving them. The fact that 90% of the items have plastics components may either suggest the work is indestructible or lead to a sense that it will deteriorate quickly. Decisions on care could even be based on the perceived value of the media: in this case low.

PROTOCOLS AND PRACTICE

Any object acquired by a museum has to meet specific criteria relating to the particular museum's collecting policy. BMT's Collecting Policy 2014–2019 states that the collection holds: 'modern and contemporary art, particularly British art, abstract painting and work by over 25 international contemporary artists on the theme of the metropolis and urban environment' (BMT Collecting Policy 2014). *ARTicle 14* thus provides a work that is squarely within its collecting policy. Tate's policy is wider encompassing British art from circa 1500 and international modern and contemporary art from circa 1900 to the present. It seeks to represent significant developments in art in all areas covered by this remit and in particular to collect works of art that are of outstanding quality as well as works that are of distinctive aesthetic character or importance. Tate also states that the decision-making process must take into account the condition of the work, and the costs of conserving it and making it ready for potential display (Tate Acquisitions Policy 2017). Tate's remit is contemporary art but there is also a current trend among other national, civic and local council collections to focus on this area, including BMT. The question such an acquisition as *ARTicle 14* raises is, are staff members equipped to manage this new area of collecting, which has similar but subtly different needs to historic artefact collections. Certainly, it has proved a costly acquisition in terms of both its financial cost and the time its care has required. The extensive range of materials that artists are using, including especially plastics, ephemera and found items (also often made of plastics) and the involvement of the artist or artist's estate in the artwork's ongoing future calls for a different approach. Additionally, complex installation art requires detailed understanding of how the different materials may interact together and within the environment, not to mention the potential interaction of the public.

Normally, during the acquisition procedure, museum staff develop a case for collecting the artwork, provide a detailed provenance, and consider issues surrounding its public display and condition. In the case of a contemporary artwork such as this, a 'living artist interview' is also carried out, with the aim of understanding the intention for the artist and the level of change acceptable in perpetuity in the museum's care. In the case of *ARTicle 14*, in spite of its complexity from the conservation perspective, not all the prescribed questions relating to the 'condition of the work, and the costs for conserving and the potential for display' took place. As the artwork was acquired at the point of de-installation from a loan exhibition, it was fast-tracked through the acquisition procedures. The conservation team only became involved when the challenges its many materials posed, in relation to appropriate storage to ensure its longevity and re-display, became clear.

This oversight demonstrates how this significant acquisition was from its beginnings in the museum context treated differently from other significant but more traditionally constructed works of art. While curators, concerned about its message saw it simply as a single unit, conservators, prioritising its long-term preservation saw it as a complex assemblage leading to a multitude of questions:

- How do I record an artwork made of 711 pieces? Where do I start?
- Many of the items are made of plastics. I do not know what type or how to deal with them. How will I categorise them on the documentation system?
- It took the artist and technical team a week to install. How am I going to de-install and document it?
- There appears to be some debate on how to (or even if to) dismantle the cart?
- These items are second hand and damaged already, do I need to record that? How detailed should the record be?
- Will my decisions put the integrity of the artwork at risk? How do I balance integrity with longevity and resource?

Condition reporting of museum objects has become simplified over the years. Reams of text that had to be interpreted and matched to the object have been replaced by digital images and simple annotations. When it comes to contemporary art, there are nonetheless often challenges: non-standard methods of display; the artwork arriving with limited documentation on condition and installation; and issues around maintaining the artist's vision and original aims for the artwork. Additionally, as noted by teams at the V&A (Nodding and Egan 2011), and especially relevant to *ARTicle 14*, a key query is what is actually considered part of the artwork, and what is supporting equipment: are the wires holding the items in place, artwork or support? Is the sticky tape part of the artwork or a vehicle for adhesion? The frame of the cart, with hand rail, wheels and break stops are considered support by some curators, literally the vehicle for the artwork; however, conservators view everything as part of the artwork making it an unusually complex entity to understand and manage.

The V&A's approach is to have, in addition to an 'artist interview', an 'artist condition statement'. Tate receives installations guides from the artist or in some cases creates them with the artist. These note all material type/components, concerns relating to light exposure, a maintenance schedule, expected operational and visitor wear, play observations relating to TVs, DVDs and so on, and handling guidelines including packing and unpacking (Nodding and Egan 2011). However, even such detailed document can be open to misunderstanding. *ARTicle 14* has an artist's interview (Jo Digger, New Art Gallery Walsall, 26 March 2013, Birmingham Museums Trusts acquisition files); however, responses to the standard questions have been found to be open to different interpretation by different museum professionals.

SOURCES AND IMPACT OF BIAS

It has been demonstrated that artworks made of plastics or with significant plastics components do not always receive the same attention at point of acquisition as artworks made of more traditional materials. Analysis of the acquisition process undertaken for *ARTicle 14* suggests that this may have arisen partly because of bias. For example, respect for the artist appears to have led to insufficient questioning of his statements:

Museum Q6:   Do the objects have to be in exactly the same place each time we install it? Are some changes acceptable?

Artist: Yes, small changes acceptable. Wired pieces can stay on.

If the artwork does not need to be exactly the same on each re-assembly, can the level of documentation be reduced? Potentially this could save time and resources. The artist did not design items to stay in situ, but has subsequently offered this option as a time-saving exercise. This sets up a conflict in the conservator's mind. Leaving items in situ puts stress on the wire and the area to which the object is attached. The whole cart frame could be dismantled but this cannot happen with items wired on. This changes the whole way the item can be transported and stored, adding extra decision-making and potential compromises in the care of the art-work as well as cost to its transportation.

Interpretation of information in a way that confirms the questioner's own preconceptions is another form of bias that can be misleading. In the artist interview, there is a statement before question 5 stating: 'some of the contents of *ARTicle 14* will deteriorate relatively quickly (in museum terms) as some of them are made of unstable medium'. The conservator's question to the curator was, 'which materials were you thinking of'? 'Plastic' was the response. The conservator, however, was thinking about the audiovisual content, the unframed photographs and the cardboard. They understood and could see that some of the plastics were damaged or aged and their questions were 'which plastics are involved in the artwork' but not that they would be the first to deteriorate. Within collections, plastics and rubbers often display visible deterioration within two to twenty-five years of acquisition (Shashoua 2008, 52). But what informa-tion do we have to tell us this is the case with these objects? There is no qualified evidence to back up this assumption, such as which plastics mate-rials are involved, or why we consider them unstable. However, given that the artist wanted deteriorated objects to be repaired not replaced, it was all the more important to have the specific plastics identified. Only then could we know if potential treatments were available.

Three possible solutions were offered to question 5:

Museum Q. 5a:	Do you want to leave the deterioration as part of the artwork?
Artist A:	No. You can change things slightly. It should look like a proper cart with saleable items on it not like an old deteriorating object that no one wants to buy from.

Hazoumè perceives these objects to be in good condition. However, as they have already been discarded by the 'West', the European visitor may not have the same opinion. They may already see it as a cart of 'rubbish'. To a conservator, items are already deteriorated and will continue to do so, so the question is, what is the acceptable level of deterioration?

Museum Q. 5b: We could try to replace the deteriorated object with an equivalent and very similar object.

Artist A: Yes. Only if something is stolen then replace it with similar.

What does 'similar' mean? Is it acceptable to replace plastics toys with metal toys? 'Only if it is stolen' is limiting to the conservator, to be able to replace deteriorating plastics may be the only option if treatment is not viable.

Museum Q. 5c: Or we could carry out conservation work on the object to try and keep it intact as long as possible.

Artist A: Yes, repair things.

In spite of the significant collector of contemporary art, Charles Saatchi's view: 'There's a squad of conservators out there to look after anything an artist decides is art' (Thompson 2012, 93), this is a lot to ask of a conservator, particularly if the materials have no identification other than 'plastic'. With the answers we are confirming what we think we know, rather than questioning what we do not understand about the materials and therefore the deterioration of this artwork.

Eventually, using infrared (IR) spectroscopy, the plastics objects were analysed (Katherine Curran, Analysis of Plastic Materials in Article 14, by Romuald Hazoumè, Institute for Sustainable Heritage, University College London December 2015. Unpublished) and found to include nine main types: polystyrene (PS); acrylonitrile butadiene styrene (ABS); polycarbonate (PP); polyvinyl chloride (PVC); polyethylene terephthalate (PET); polyurethane foams (PU); polypropylene (PP) and nylon. Of particular significance was the use of solid-phase micro-extraction followed by gas chromatography-mass spectrometry to determine the volatile compounds off gassing from particular objects/materials (SPME- GC/MS) because it enabled identification of deterioration in specific pieces. For example, the PET-based magnetic tape showed organic acid off gassing suggesting the start of the deterioration process, as did both Barbie dolls. The cards they were fixed to were also starting to deteriorate. This information helped to inform conservation practice and research into how best to manage these plastics and how to store the items. In this case, the solution was to separate the dolls from the card supports when not on display.

Scientific analysis of the materials also improves documentation. Although museums and galleries always document the particular metal or wood of which an object is made, curiously practice with plastics is different. To take Tate as an example, if the particular plastics is known, it is recorded but otherwise the artwork is listed simply as 'plastic'. This is, however, for artworks that are wholly made of plastics or have a major plastics component; if an artwork has, for example, a piece of PVC piping as part of its mechanism, it is much less likely to be recorded. This is problematic as different plastics require different optimal conditions to prolong their lives as long as possible. The question is, why do museums and galleries tend to record plastics in less detail than other materials although knowing which plastic has more significance in terms of how they are stored and thus more bearing on their condition and longevity?

This may stem from cultural bias. The statement 'high volume automated production of plastic objects has allowed for social and economic growth, but once outdated can be seen as valueless flotsam and jetsam of mass culture' (Fiell and Fiell 2009, 7) reflects some of Hazoumè's intentions. However, do these perceptions of the contents of the artwork also have an impact on how it is treated within the museum context?

As the artwork is made up of second-hand goods on a market cart and set in a market scene, it can easily be associated with cheapness, influencing a perception of the artwork as figuratively less valuable and less important (Dobelli 2014, 48). A high percentage of viewers said, I had one of those, I threw it away, making an association with the worthlessness of the items. Additionally, the way the objects are displayed on it, gives the appearance that not much thought has gone into it or that it is not meant to hold for that long, so it is alright if bits drop off. The plastics items are seen as 'cheap things for sale', dead social history, rubbish tip items, with the result that the association distracts or blurs the artist's message and intent. That, in turn, may lead to more risks being taken with the artwork and less thought being put into its care.

Contemporary art can be difficult to define. As discussed in Chap. 3, artist Jeff Koons moved a vacuum cleaner from a department store to the gallery re-contextualising the object and challenging the viewers' perception (Thompson 2012, 87). The lines between art and everyday objects are further blurred by the context in which some objects are displayed in museums and galleries, for instance, the same Louis Vuitton bag has been displayed in the V&A as design, in the Guggenheim as art and in the shop as a commodity (Thompson 2012, 11). Cultural bias suggests a raft of questions in relation to *ARTicle 14*. Does its plastics materials and objects have these blurred branding issues? Does this affect the way both visitors and staff perceive the artwork? Is there too much of the everyday in these objects for their elevation to art? Are the objects so known to us and tactile that one cannot help but want to interact with the artwork? Is there an unconscious hierarchy of materials value? If the objects in *ARTicle 14* were gold-plated would museum staff think differently about the artwork? The answer to this question is probably yes, certainly gold is particularly attractive to some members of the public: whilst the artwork was on display two gold-coloured metal watches were stolen. This suggests that there is a hierarchy of materials and plastics are not at the top of this list for some.

CONCLUSION

This chapter has explored attitudes to plastics at the V&A and found that, although they are collected, they are not valued as a coherent materials group worthy of study for their own sakes as is the case with paper, textiles, furniture (woodwork), ceramics and metalwork. There is no gallery of plastics that presents their history and use, as there are for these other materials. However, objects made of plastics are valued if they make a specific contribution to design's broader trajectory.

The chapter has also demonstrated, through a focus on a particular artwork, that where plastics are involved, museum protocols may not always be applied with the same rigidity as is usually the case, leading to the question: is plastics' prejudice prevalent even among curators and conservators? Reflection on these discussions suggests that people, including professional museum staff, are affected by unconscious bias seeing plastics as cheap, familiar, challenging (as art), replaceable, difficult to repair, stable, and conversely, that they are unstable. From the museum professional standpoint, the consequences are that we may not know how to ask the artist the best questions to ensure the best possible interpretation and longevity of use for the object.

NOTES

1. In 2015, when the conference on which this book is based took place, Cane was Conservation Manager at BMT.
2. The artwork's title refers to the rumour that all African constitutions contain a hypothetical final article which translates as 'When all else fails, do what you need to do, because no-one else is looking out for you!'

BIBLIOGRAPHY

BMT Collection Development Policy 2014–19. https://www.birminghammuseums.org.uk/about/our-organisation/policies-plans-and-reports. Accessed 20 April 2020.

Breward, C., and Ghislaine Wood, eds. 2012. *British Design from 1948: Innovation in the Modern Age.* London: V&A Publishing.

Burton, Anthony. 1999. *Vision and Accident, the Story of the Victoria & Albert Museum.* London: V&A Publications.

Cascini, G., and P. Rissone. 2004. Plastics Design: Integrating TRIZ Creativity and Semantic Knowledge Portals. *Journal of Engineering Design* 15 (4): 405–424.

Dobelli, R. 2014. *The Art of Thinking Clearly.* London: Sceptre.

Donoghue, Katy. 2019. Romuald Hazoumè's Masks Reveal the True Face of an Individual. https://www.whitewall.art/art/romuald-hazoumes-masks-reveal-the-true-face-individual. Accessed 20 April 2020.

Dorotheum. 2016. Pratt Chair. https://www.dorotheum.com/en/l/1718743/. Accessed 20 April 2020.

Ferriani, B., and M. Pugliese. 2013. *Ephemeral Monuments.* Los Angeles: J. Paul Getty Trust.

Fiell, C., and P. Fiell. 2009. *Plastic Dreams.* London: Fiell Publishing Limited.

Griffith, Roger. 1996. Two Pooped-Out Pop Chairs: What Is the Future for Our Plastic Collections? *V&A Conservation Journal* 21. http://www.vam.ac.uk/content/journals/conservation-journal/issue-21/two-pooped-out-pop-chairs-what-is-the-future-for-our-plastic-collections/. Accessed 20 April 2020.

Hvattum, M., and C. Hermansen, eds. 2004. *Tracing Modernity; Manifestations of the Modern in Architecture and the City,* 128–129. London: Routledge.

Keneghan, Brenda. 1996. Plastic? Not in My Collection. *V&A Conservation Journal* 21. http://www.vam.ac.uk/content/journals/conservation-journal/issue-21/plastics-not-in-my-collection/. Accessed 20 April 2020.

Lambert, Susan. 2012. Plastics – Why not? A Perspective from the Museum of Design. In *Plastics in Extreme Collecting, Challenging Practices for 21st Century Museums,* ed. Graeme Were and J.C.H. King, 168–180. New York and Oxford: Bhergahn Books.

MoMA. The Collection. n.d.. https://www.moma.org/collection/works/87278 ?classifications=any&date_begin=Pre-1850&date_end=2020&locale=en&pag e=1&q=Pesce&with_images=1. Accessed 20 April 2020.

Mundy, Jennifer. 2020. Lost Art: Naum Gabo. https://www.tate.org.uk/art/artists/naum-gabo-1137/lost-art-naum-gabo. Accessed 20 April 2020.

Nodding, H., and L. Egan. 2011. The Contemporary Art of Documentation. *V&A Conservation Journal* 59. http://www.vam.ac.uk/content/journals/conservation-journal/spring-2011-issue-59/the-contemporary-art-of-documentation. Accessed 20 April 2020.

October Gallery. n.d. Romuald Hazoumè. http://www.octobergallery.co.uk/artists/hazoume/. Accessed 20 April 2020.

Shashoua, Y. 2008. *Conservation of Plastics*. Oxford: Elsevier.

Tate Collection and Disposal Policy 2017–2020. https://www.tate.org.uk/about-us/policies-and-procedures. Accessed 20 April 2020.

Thompson, D. 2012. *The $12 Million Stuffed Shark*. London: Aurum Press Ltd.

V&A. n.d.-a. Search the Collections. http://collections.vam.ac.uk/item/O146046/vacuum-cleaner-dyson-james/.

———. n.d.-b. Search the Collections. http://collections.vam.ac.uk/item/O142142/armchair-eames-charles/. Accessed 20 April 2020.

———. n.d.-c. Search the Collections. http://collections.vam.ac.uk/item/O112351/blow-armchair-de-pas-jonathan/. Accessed 20 April 2020.

———. n.d.-d. Search the Collections. http://collections.vam.ac.uk/item/O1227039/pratt-chair-pratt-chair-pesce-gaetano/. Accessed 20 April 2020.

———. n.d.-e. Search the Collections. http://collections.vam.ac.uk/item/O1299501/materialized-sketch-chair-front-design/. Accessed 20 April 2020.

———. n.d.-f. Search the Collections. http://collections.vam.ac.uk/item/O114267/rcp2-chair-chair-atfield-jane/. Accessed 20 April 2020.

———. n.d.-g. Search the Collections. http://collections.vam.ac.uk/item/O101947/chair-bar-knell/. Accessed 20 April 2020.

Vitra Design Museum. n.d. Gaetano Pesce. http://collection.design-museum.de/#/en/person/7138?_k=hgl24a. Accessed 20 April 2020.

Williams, Gareth. 1996. Plastic, Pop and Mass-Produced Design in the V&A's Collections. *V&A Conservation Journal* 21. http://www.vam.ac.uk/content/journals/conservation-journal/issue-21/plastic,-pop-and-mass-produced-design-in-the-v-and-as-collections/. Accessed 20 April 2020.

CHAPTER 14

Materiality and Perception: Plastics as Precious Materials

Gerson Lessa

This chapter introduces research that cross-relates human perception, with all its subjectivity, to quantitative assessments of material properties. Its aim is to understand better the widely varying perceived values of plastics materials in the form of finished products.

As claimed by Friedel:

> [A] history from things also should begin with the materials that go into our artifacts. [It] is ironic that studies of material culture should so neglect the actual materials that go into creating culture. [...] The form tells us what the object is [...] but it is only when we take into consideration the materials of each object that we can begin to appreciate the real history of each. The material itself conveys messages, metaphorical and otherwise, about the objects and their place in a culture. (Friedel 1993, 42)

Materiality is a fundamental aspect for the analysis of material culture and plays an especially important role in the existence of goods made of plastics. Given plastics' ubiquity, relatively recent history and the not

G. Lessa (✉)
Federal University of Rio de Janiero, Rio de Janiero, Brazil
e-mail: Gerson.lessa@eba.ufrj.br

© The Author(s), under exclusive license to Springer Nature Switzerland AG 2020
S. Lambert (ed.), *Provocative Plastics*,
https://doi.org/10.1007/978-3-030-55882-6_14

always positive perceptions they engender, quality goods made of the best plastics materials had for a long time to stress the positive values of their materiality in order not to be seen by consumers in a depreciative way. This perception is still alive today. Thus, it may still seem provocative to classify plastics as precious.

However, long gone are the days when the plastics industry had to prove the practical qualities of its materials and struggle for their acceptance. Viewed first as novelty materials, then as surrogates, especially at times of shortages as in wartime, plastics have developed into a vast array of specialised materials with visible or concealed applications that the whole industrial world could hardly afford to do without. Today the plastics industry is capable of consistent designs and high-quality mouldings that can hardly be associated with cheapness or bad quality. Even so, the lesser fringes of the consumer goods market manage to sustain the image of shoddiness which plastics were associated with in their earlier days. Thus, quality perception is still an important issue to be addressed by designers when considering the use of plastics.

THE MATERIALITY OF PLASTICS

Central to the understanding of the value perception of plastics is awareness of the qualitative differentiation between the many different plastics materials. Of special importance are plastics developed for unusual applications and requiring unusual production and commercial strategies. In the short period from the beginnings of industrial plastics production in the mid-nineteenth century until their exponential growth for commercial purposes after World War II, their use to substitute for and mimic traditional materials harmed their public image. Plastics began to be viewed as cheap and shoddy substitutes in opposition to wholesome natural substances, despite the ever-growing practical and useful number of applications to which these new materials were being put.

A good example of a plastics material capable of convincing imitation is Bois Durci. Developed in France in the mid-nineteenth century. It is a composite of natural substances: animal blood, serum and sawdust. These ingredients were moulded under heat and pressure into objects that closely resemble quality carved wood. The significant step it made possible was the transfer of production of intricately decorated hand-carved objects to mechanical reproduction. The resulting material looks very much like wood. Initially, the colouring agent was sawdust itself, with the animal polymeric protein binding the particles together, resulting in a very sturdy material with perfectly polished surfaces capable of reproducing the finest detail. This material and the production of objects moulded from it provide a typical example of the successful transition from hand-carved to moulded; from artisanal to mechanical; from natural to artificial; from tradition to innovation.

Although early plastics played a significant role in the development of machine production, the fact that this also involved the imitation of craft practices and natural materials is key to understanding the classic negative perception of plastics, especially since not all plastics materials achieved such convincing imitations. One proclivity of these new means of production was the predominant economic logic that the cheaper the production costs the better, even if detrimental to quality. The poor design and low-quality mouldings with thin walls and cheap finishes of fragile plastics goods produced in their thousands became a formula that marred the reputation of an entire class of materials.

The way in which materials were used was a major determinant in these perceptions. Cellulose nitrate, or celluloid as it was called in the United States, for example, was capable of good and bad imitations. It was the first material that made possible the production of thin sheets that could then be mass thermoformed or blow moulded into many different shapes. When it was used to simulate heavy materials like stones, onyx or jade, for instance, the end products were often extremely lightweight, warm to touch and easily dented, the opposite of the materials they were pretending to be. However, being a versatile material, it could also be used to mould solid objects. The increased mass added weight and hardness, making the resulting objects resemble more closely the materials they were trying to replace, completely changing how they were perceived. Some of the best quality imitations, like decorative curios made of ivory-looking cellulose nitrate made in China in the early twentieth century, are still confusing sellers and buyers today (Fig. 14.1).

However, more generally, the utopian and fantastic rhetoric of plastics advertising was at odds with the predominant reality of cheap and poorly made products. Reality proved more persuasive than fantasy to the detriment of public perception of plastics.

A proper grasp of the concept of materiality requires a high awareness of the properties of the vast range of plastics now available. However, although these properties arise from their chemical composition and molecular structures, the interaction between an object and a consumer is qualitative rather than quantitative, arising from how contact with the

Fig. 14.1 Celluloid boxes mimicking natural materials. (Photo: Gerson Lessa)

object is perceived by at least four of our five senses: touch, sight, hearing and smell. Contact with an object allows us to compare fundamental elements of materiality: mass, density, stiffness, flexibility, softness, temperature, transparency or opacity, colour, superficial finish and even the sounds it makes and how it smells. A well-trained individual may be able to identify the most common plastics by simply manipulating them and sometimes, even, by just looking at them from a distance. Similarly, consumers will be influenced consciously or not by these qualities, judging the value, significance and relevance of an object through their emotional and associative responses to its design. Material qualities affect the design by the way materials are applied and are, thus, responsible in great part for whether our interaction with products is positive or not.

TECHNOLOGICAL DETERMINANTS AND PERCEPTION

Baudrillard has pointed out how distanced we are from the technology that creates objects:

> The technological plane is an abstraction: in ordinary life we are practically unconscious of the technological reality of objects. Yet this abstraction is profoundly real: it is what governs all radical transformations of our environment. It is even (…) the most concrete aspect of the object, for technological development is synonymous with objective structural evolution. (…) It is starting from this [technological] language, from this consistency of the technical model, that we can reach an understanding of what happens to objects by virtue of their being produced and consumed, possessed and personalized. (Baudrillard 2002, 5)

Design and production in plastics have achieved much more than imitation. The versatile physical properties of cellulose nitrate and the many processes it can undergo made it suitable for imitation. Other properties, for example, its fragility and flammability, made certain uses unviable. In particular, they restricted its potential in structural applications. However, in the form of clear film, it contributed to the progress of photography and the birth of the movie industry and thus to a significant technological breakthrough, as many other plastics have done since.

After World War II, when many new plastics materials with different capabilities became available for civilian use, plastics found new ways to challenge the use of well-established materials and develop new industries and markets. Experimentation was rife. A wide typological range of products was produced in plastics and consumers found themselves judges of the results. Tough new plastics, like melamine formaldehyde, were widely introduced for tableware, with a matching marketing strategy that promoted their positive characteristics: especially their toughness in contrast to breakable china. However, the shiny hardness of glazed ceramics could not be matched by melamine ware, which scratched, lost its shine and became stained in normal use. It also cracked if the water was too hot. In spite of how the material was marketed, it was consumers that decided the occasions for which melamine or china were better suited, with the result that melamine ware was mainly used for informal daily meals or relegated to picnic ware. Top quality objects, being pleasingly heavy and well moulded, look and feel satisfying to the touch. However, the signs of wear, which quickly appear, completely undermine the material's favourable first impressions. Ceramics still won the day for formal occasions.

Other plastics, including thermoplastics like polystyrene and polyethylene, were tentatively used for similar applications. The American magazine *Consumer Reports* specialised in laboratory testing and rating all sorts of consumer goods as a guide to buyers, including dishes made of polystyrene, urea formaldehyde and melamine. They were not impressed with the results, claiming the objects failed to fulfil the promises of the makers: 'Plastic dishes are, of course, far harder to break than ceramics. It is unfortunate that in most other respects they do not perform so well as their ceramic counterparts. Plastic tableware is much more easily damaged by heat than most ceramics; it is greatly weakened by sudden and severe changes in temperature; it is fairly scratched, and generally stained by some common foods... It lacks the hard glaze of ceramics'. Along with melamine ware, polystyrene pieces were also tested and it was found that they were, in all respects, less suitable to the proposed tableware use, being easily broken and prone to stressing and deformation when exposed to heat (Consumer Union of US 1951, 8–10) (Fig. 14.2).

Despite the disappointing results of these tests, the magazine ran the tests again a few years later, this time centered on melamine ware only, with more optimistic results: 'More brands are available, in more shapes, colors and designs than ever before—more of them rank high in overall quality ... Consumer acceptance of plastic dinnerware is now such that about 20% of all the domestic dinnerware sold [in the USA], according to industry estimates, is plastic. This acceptance is largely attributable to the use of a single plastic, melamine-formaldehyde' (Consumer Union of US 1954, 536–538).

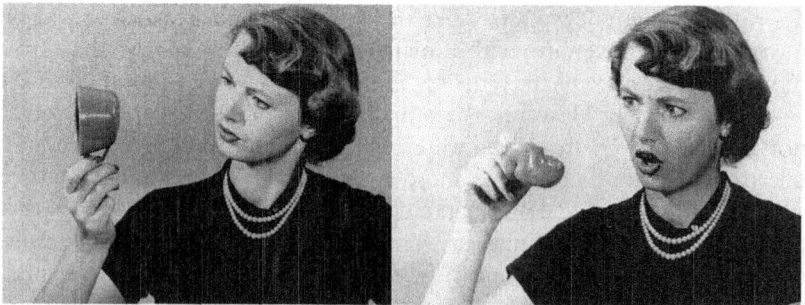

Fig. 14.2 Failed plastics: The results of sterilising a polystyrene cup in boiling water. (Photos: Consumer Reports Magazine, 1951)

It is clear that, despite these tentative uses of the new plastics materials group in established consumer niches like tableware, the simple possibility of actually moulding the object did not of itself mean it would perform well or gain acceptance. It is the sum of physical properties, well managed by design, which determine the practical and commercial success of an object and ultimately, its usefulness. Although this sounds quite obvious today, it seems not to have been so obvious in the context of the fast-growing plastics industry of the mid-twentieth century.

Plastics materials can be perceived perfectly to embody Miller's concept of *meta-commodities*: materials or goods that are perceived for pragmatic or ideological reasons as embodying a series of capitalist and industrialist threats (Miller 1998, 169–170). The immediate association of plastics with environmental pollution is the most obvious, but not only, example. Consumerism, planned obsolescence, the banalisation of mass-produced goods, the natural/artificial opposition, and the common association of plastics with kitsch, have all contributed to a well-established and strong opposition to plastics goods. For this, the industry has nobody but itself to blame.

In relation to industrial goods, a symbolic circuit can be identified in which generic interpretations of the status of an artifact are given through the 'material/object relation' and the 'consumer/object interaction' (Shove et al. 2007, 111). The basic narrative structure is one in which objects speak of materials and material discourses affect how objects gain and lose symbolic meaning. My consideration of the materiality of plastics fits perfectly in such a perspective. A good example of these material effects and evaluations can be found in two well-known polyethylene uses: the good, plain, functional and permanent design of the best Tupperware bowls is a far cry from the generally despised shopping bag, where permanence is negative: too short as a throwaway object and too long as an environmental hazard.

When design and production standards are set high, the use of thermoset plastics in the material/object relation may result in more successful objects in the consumer/object interaction. That is because the physical properties of most thermosets are perceived differently from those of most thermoplastics: the latter being less dense, softer, warmer and inclined to make less noise than most thermosets (Fig. 14.3).

Nonetheless, material aspects like mass and density can, sometimes, compensate one for another: a larger mass of a light material may bring an object to be perceived more favourably than one made with less mass of

Fig. 14.3   Tupperware cup, made of polyethylene thermoplastic and Color-Flyte cup, made of melamine thermoset plastic. (Photo: Gerson Lessa)

the same material. This is true also of objects made of other materials. Both the melamine tableware and the cellulose nitrate curios are examples of products with greater material mass making them not only more resistant to breakage but more rewarding to the senses.

The density properties of the most commonly available thermoset plastics in the 1940s, casein formaldehyde, phenol formaldehyde, urea formaldehyde and melamine formaldehyde, are all found in the range of 1.3–1.87 of a scale that goes from 0.9 to over 1.9 grams per cubic centimetre. Only two widely used thermoplastics, cellulose nitrate (celluloid) and polyvinyl chloride (PVC) can be found in the same range. Other widely used thermoplastics, for example, polyethylene, polystyrene, nylon and cellulose acetate, are less dense and found in the 0.9–1.3 range (Lima 2006, 147–177; Lefteri 2008, 272–277). Ideally, the binomial mass versus density must be taken into account in an object's design in terms not only of economy of material but also in how both its use and sense of permanence will be affected. However, density is just one of the many characteristics to be considered in a quantitative assessment of materiality; others, like scratch hardness, that measure the resistance of a material's surface, will also affect the value perception of both materials and the goods made from them.

The ubiquity of cheap and disposable plastics conflicts with concepts such as rarity and permanence embedded in most cultures as attributes of the highly valued, the precious. There are, though, some interesting niches today in which these relationships may be turned upside down and where we can find certain plastics goods valued to astonishing degrees.

CAST PHENOLIC RESINS

The effect of materiality on our appreciation of materials is clearer when examined in relation to a special case where high value is now attributed to some plastics objects in the marginal markets of vintage goods, especially those made of cast phenolic resins.

Phenolic resins were the result of empirical experiments at the turn of the twentieth century, a time when a true understanding of polymer structures and chemistry as a whole was still in its infancy. Leo Baekeland, like many other independent researchers, was looking for materials of industrial interest in his own laboratory when in 1907 he made a successful synthesis of a resin-like material that was not like any other substance obtained in nature. It was the result of the polymerisation of two organic substances, phenol and formaldehyde, resulting in the first synthetic polymeric substance. It was a hard resin with interesting properties, the value of which for many industrial applications, was quickly recognised. These include perfect mouldability, hardness, dimensional stability, electrical insulation and chemical resistance. They were at first associated with functional inner parts of consumer goods or industrial machinery but, in order to be useful in these applications, this hard but otherwise brittle resin had to be given structural strength. This was achieved by the addition of fillers, leading to composite materials. They were obtained by soaking paper sheets in the resin to achieve laminated materials that could be cut and machined to their final forms, as in Formica, or by blending ground resin with wood flour or other fillings to obtain a powder mouldable by heat and pressure, a technique that had also been used previously with natural plastics like horn, Bois Durci and Shellac. These filled resins were very successful and quickly found their 'thousand uses', as promised in advertisements for Bakelite. Thus, they began also to be used to mould exterior casings of domestic appliances and small functional objects like handles or shavers: applications that were meant to be visible and to be the physical interfaces between the object and the consumer.

Despite the aesthetic and functional value of these moulding powders, their dark colour and the interference of the fillers made effects like transparency or pale colours impossible. Thus the material is associated with a limited range of colours: black, shades of brown, red, ochre, green and blue. However, even these colours were not permanent and tended to darken or change completely with time, with the result that, today, most old mouldings that are not black or red, tend to be found in shades of murky brown. This was an obvious limitation to the use of the material.

Subsequent formulations of phenolic resins resulted in water clear variations known through trade names, such as Catalin or Marblette, which came on the market in the early 1930s. Unlike the functional moulding powders, which were so useful for industrial purposes, they could be cast in colourful opaque, translucent, transparent, clear and mottled colours and were valued rather for their aesthetic and tactile qualities. They were produced and sold in precast standard shapes such as rods, tubes, sheets and slabs aimed at manufacture in an artisanal fashion of small objects by cutting, slicing, machining and polishing for the novelty industry or home and school workshop (Lockrey 1940) (Fig. 14.4).

They were also used to cast specific proprietary designs, the most relevant to this study being radio cabinets, which also involved hand finishing of varying degrees of complexity.

A Matter of Re-Evaluation

Many industrial objects turn into desirable collectables with the passage of time. It is commonplace now to see what were everyday products designed by famous designers on auction websites selling for final prices of four or five digits. This happens also with products made of plastics, sometimes even when no famous names or brands are attached. This is particularly the case with cast phenolic products but the value given to them is not linked to their material characteristics alone. Their form and style also have a decisive influence on how they are re-evaluated. General style signifiers like art-deco, streamlining, machine age and space age are important to collectors, being representative of popular period-defining forms as well as being attributes that make it easier to identify and date the pieces concerned.

Broadly speaking, when an outstanding design is made of any outstanding material, the more desirable it will be to both its original consumers and to subsequent collectors. Objects made of cast phenolics in particular often fetch high prices but what makes these cast materials so special? There are two reasons, one material and the other process related.

The material-related reason is attributable to their high density, their hard and lustrous surfaces and the depth of their colours and mottling. The high density of the material gives a feeling of solidity and stableness. As demonstrated above, admiration grows in direct proportion to the

Fig. 14.4   Precast phenolic shapes (upper row) and objects machined from them, c.1940. (Photo: Bakelite Corporation)

mass of the object. Colour, however, is a more complex subject because, as with filled phenolic resins, cast phenolics tend to discolour, certainly when exposed to light but also even when they have been protected. The colours, except in the case of black and red, which we see these products made in today are seldom their original ones. White, which was common at the time of production, can no longer be found: originally white products have turned to different shades of burnt yellow or orange, a colour commonly called butterscotch. Water clear phenolic also has not survived. All products made with this material have turned to a rich golden yellow, also known as apple juice. Although these are not the original colours, they are nonetheless pleasing. Blues, pinks and greens have, however, changed to sometimes very unattractive colours, many of them simply going a murky brown. Mottled phenolic was mostly made by mixing opaque white pigmented and transparent pigmented resins, just before pouring them into the moulds. So most mottled phenolics are found today as mixes of butterscotch and a second colour. The most valuable materials are those that have survived the discolouring process well, usually quieter but still attractive hues. The resulting colour palette is limited but nonetheless distinctive and easily spotted by connoisseurs.

The process-related reasons are attributable to the work applied to the precast forms, their design, production and finish. The fact that most cast phenolic objects are made with a considerable degree of handwork makes each one unique. The informed collector can see the indicial clues of the artisan's handiwork and the equipment used: the more complex, the more desirable the object. The casting/machining processes leave telltale marks and imperfections that are very different from those left by more common plastics moulding processes, like injection moulding. This makes the products resemble even more closely the traditionally carved materials phenolics may have wished to emulate. Some of these imitations are still mistaken for natural materials, for example, amber. However, it is only as a result of a striking confluence of design, material, colour and finish, and after the slow maturing of time that phenolic plastics come to be deemed especially precious.

Precious Plastics

Two types of objects made of cast phenolics attract most attention from collectors and are, therefore, particularly representative of the value criteria under discussion. These are radios and bracelets. Their iconicity is demonstrated by the substantial prices outstanding pieces fetch on the market. They are, however, niche markets, each with very different cultural rationales and backgrounds.

Collectors' interest in cast phenolic radios is based on three factors: their place in the history of radio technology, their representativeness of their period style; and the materials used in their manufacture. From crystal sets to transistors, an immense range of designs in many materials is available. When radio technology became widespread in the 1920s, the growth in demand allowed fabricators to move from wooden cabinets to mass production in plastics composites, like Bakelite, to produce highly finished products in simple and fast industrial cycles. By 1937, the first cast phenolic cabinets were being produced in the United States. The complexities of the artisanal casting processes restricted the use of the material to special limited runs of gaudy and expensive radios (Hawes 1996). They have the same place today in the collectables market. Harder to come by, especially when in attractive colour combinations or mottling, they may fetch impressive prices when compared to mostly black or brown examples moulded in Bakelite composites (Fig. 14.5).

Bracelets and bangles from the 1930s and 1940s are much sought after by jewellery and plastics' collectors alike. Unfortunately, however, information about their production is scarce. They are seldom marked or signed which makes their precise dating difficult. The origin of specific designs is also hard or impossible to trace. Their design varies greatly. Thin and plain bracelets, sometimes referred to as spacers, are only valuable

Fig. 14.5 Emerson AU-190 radio as sold on eBay, 2015. Auction page no longer available. (Screen Capture: Gerson Lessa)

when colourful or found in sets of identical shape and colour. The heavier a simple, uncarved bangle, the higher its value. This is because the material qualities of phenolics are amplified and shown more clearly: the simplicity of form is compensated for by bigger areas of mottling and greater mass. When more handwork and complexity of design are added, value rises accordingly. Faceting, deep carving, lamination, geometric modernistic styling and intricate floral designs, all add desirability and thus associative and financial value.

At the time of manufacture, unlike the radios, they were not aimed at the wealthy. Most were sold at dime stores although some more elaborate sets did find their way into department stores. Nonetheless, they never cost anywhere near as much as the prices collectors will pay for them today (Davidov and Dawes 1988). When a famous provenance is added, like in an example from the Andy Warhol collection, prices may skyrocket. As of July 2015, the website of Skinner Auctioneers showed a big and heavily carved bracelet originally from his estate that was auctioned in 2005 for US $6756 (Skinner Inc n.d.) (Fig. 14.6).

Although cast phenolic objects made in Europe exist, production of the raw materials, the precast forms and the jewellery took place mostly in the United States. Other plastics were used in different countries for some of the applications we find in cast phenolics. An example is casein formaldehyde, generally referred to as Galalith, produced from cows' milk in countries including the United Kingdom and Brazil. Being very brittle and prone to cracking and distortion because of its too easy absorption or loss of water, it can only be thermoformed to a limited extent and was, thus, mainly worked by machining extruded rods or slabs. This semisynthetic plastic could never have been made into products like radio cabinets or bracelets as they would have been too fragile and short-lived. All this places cast phenolic as opposed to moulded phenol formaldehyde as an American phenomenon.

These objects were mostly produced in the 1930s and 1940s, and were, as in the case with most cast phenolic objects, machined from stock shapes, like cylinders, rods and sheets of varied thicknesses. The production of these shapes was complex and time consuming. The liquid resin had to be manually pigmented, which made room for endless variations, but it could not be catalysed; it hardened when baked in lead moulds inside huge kilns at 200°C for up to five days. The resulting shapes, made in clear, transparent, translucent or opaque colours, plain or mottled, were sold to small-scale workshops or hobbyists as 'the perfect workshop material' for the

SKINNER

Rare Bakelite Carved Bangle

Auction: 2270
Lot: 493
Sold for: $6,756

Auction: Fine Jewelry - 2270
Location: Boston
Date / Time : March 15, 2005 11:00AM

Description:

13/07/2015   Rare Bakelite Carved Bangle | Sale Number 2270, Lot Number 493 | Skinner Auctioneers

Rare Bakelite Carved Bangle, the wide bracelet heavily carved with palm trees and flowers, interior circumference 7 3/4 in.Note. This bracelet was part of the Andy Warhol Bakelite collection sold by Sotheby's in 1988. The original lot label from that auction is included. The bangle is pictured on p. 11 of "The Bakelite Jewelry Book". Note: Lots 421 to 543 come from the collection of Corinne "Corky" Davidov. Ms. Davidov co-authored "The Bakelite Jewelry Book" (1988) with Ginny Redington Dawes. Many lots in the collection are pictured in the book. Ms. Davidov also co-authored "The Victorian Jewelry Book" (1990) and is currently a writer for Lucire, a global fashion magazine. This collection was amassed over twenty years beginning with a few pieces owned by her mother.
Estimate $3,000-5,000

Fig. 14.6 Final bid for a Bakelite bracelet from the Andy Warhol collection, Skinner Auctioneers, 2005. (Credit: Skinner Auctioneers)

production of a vast array of small practical or decorative objects (Fig. 14.7).

The bracelets were cut to any desired length from cylinders about 30 cm tall and with diameters varying from 7.7 to 9.5 cm in diameter, with an inner diameter of about 6.5 cm. They were then turned on a lathe before being carved manually, in geometric or organic designs, by artisans using hand or power tools: saws, mills, sanders, drill presses, carving spindles and buffers (Lockrey 1940, 1–94). Thus, each piece was carved and polished by hand, and pieces with the same design show the discrepancies expected in handmade work. Other custom shapes could be cast, like radio

Fig. 14.7 Machine carved cast phenolic bracelets, USA, 1930s and 1940s. (Photo: Gerson Lessa)

cabinets, and they also required hand finishing, for example, sanding and polishing to acquire the lustrous, jewel-like surfaces for which the material is known.

Part of the current value, commercial, mythic or otherwise, attributed to objects made of cast phenolic resin lies in the fact that it is now obsolete and practically extinct. The original stock shapes and colours are no longer commercially available and the carving skills necessary for the production of new designs have become very rare. Nonetheless, as evidence of the interest still aroused by the special properties of this material, until recently companies were still producing precast shapes or specific objects made from it. An example is Raschig Co. Ltd. with factories in Germany, India and Thailand that was offering on its website at least until 2015 what was guaranteed as cast phenolic blocks and rods made with the same formula and properties of the original ones. This mention has now been removed. Although the company is still in the plastics business, production of cast phenolics has been abandoned (Raschig n.d.). As of 15 January 2018, the Belgian company Saluc S.A. was offering on their site Aramith and Preciball product ranges, including high-precision billiard balls and stock shapes made of cast phenolics. Their process is, however, hi-tech, using the latest digital technology to ensure perfect geometry, balance and performance for professional billiard players and contests (Aramith n.d.), a far cry from the traditional craft-based jewellery workshops of 1930s and 1940s America as pictured in the few books on the subject (Davidov and Dawes 1988, 79; Lockrey 1940, 1–94).

A by-product of the interest and value attributed by collectors to vintage cast phenolic jewellery is the recent emergence of counterfeits. Countless bracelets are produced in India in cast and carved resin-like polyester, which emulate the designs and the general aesthetics of cast phenolic: a threat to serious dealers and buyers, who call them 'fakelite'.

Phenolics are different from our traditional notion of precious materials like metals. They can be mixed with other scraps and melted down, so the value of the material is not affected by the shape or size and scraps have value whatever their form. Whereas phenolics, being thermosetting, cannot be melted and re-used, thus scraps are more or less worthless, and of interest only to the few contemporary artisans who rework old pieces into new objects, mostly jewellery.

Cast phenolic resins were only on the market for a short period of time for economic reasons. The processes involved could not have been more complex, time consuming and dependent on skilled handwork. They were also polluting and toxic to the workers and the machining techniques resulted in great amounts of the material turning into useless and unrecyclable waste. All in all, the process was closely related to traditional manufacturing procedures that were rapidly being abandoned by industry in favour of the new injection-mouldable thermoplastics which became available for mass production after World War II. The distinctive materiality of cast phenolic could never be a competitor in the face of such a successful manufacturing culture. However, their materiality, artisanal production, scarcity and flamboyance once formed have created a plastics product of high collectable value and desirability: a plastics material may never be as prized again.

CONCLUSION

Materials are essential actors in our world of goods: a world full of functional and symbolic roles where natural or manmade substances are used to build structures and add structural or superficial value. Thus, materials convey messages about themselves, their origin, their production, their use and their place in the cultures that make them. Plastics, although deeply associated with the values of industry, offer a wider range of messages to designers. In reflecting this plethora of physical and symbolic possibilities, results have been variable both in functional and aesthetic terms.

Material value systems change over time. This can be readily observed with plastics, which have survived their original functions and have now been reassigned as collectables. Plastics provide a somewhat novel opportunity to consider changes of value in material culture; ubiquitous materials force new considerations beyond traditional approaches related to materials with longer traditions, like precious metals. In post-consumer markets cast phenolics are judged on their designs, which typify not only a historic stylistic period but also benefit greatly from the quality of their material and the ingenuity and skills necessary to shape them: values rarely associated with plastics or any mass-produced goods.

The subtleties of plastics materiality are important not only for plotting the course of historical development but for the way we design objects with plastics materials available today. In-depth understanding of the effects of this materiality can only be fully appreciated through a complex correlation of their quantitative physical properties and the qualitative perceptions they give rise to in use. Successful evaluation of these properties may well lead to an improved capacity to forecast a more positive consumer material culture in the future.

BIBLIOGRAPHY

Aramith. n.d. Billiard Balls. https://www.aramith.com/our-quality-commitment. Accessed 14 April 2020.

Baudrillard, Jean. 2002. *The System of Objects*. London: Verso.

Consumer Union of US. 1951. *Consumer Reports*. 16:1.

———. 1954. *Consumer Reports*. 19:11.

Davidov, Corinne, and Ginny R. Dawes. 1988. *The Bakelite Jewelry Book*. New York: Abbeville.

Friedel, Robert. 1993. Some Matters of Substance. In *History from Things*, ed. Steven Lubar and W. David Kingery. Washington: The Smithsonian Institution Press.

Hawes, Robert. 1996. *Bakelite Radios*. London: The Apple Press.

Lefteri, Chris. 2008. *The Plastics Handbook*. Mies: Rotovision.

Lima, Marco Antonio M. 2006. *Introdução aos Materiais e Processos para Designers*. Rio de Janeiro: Ciência Moderna.

Lockrey, A.J. 1940. *Plastics in the School and Home Workshop*, 1–94. New York: Van Nostrand.

Miller, Daniel. 1998. Coca-Cola: A Black Sweet Drink from Trinidad. In *Material Cultures: Why Some Things Matter*, ed. Daniel Miller, 169–170. Chicago: The University of Chicago Press.

Raschig GmbH. n.d. http://www.raschig.de/index-en.php. Accessed 14 April 2020.

Shove, Elizabeth, et al. 2007. *The Design of Everyday Life*. New York: Berg.

Skinner Inc. n.d. https://www.skinnerinc.com/auctions/2270/lots/493. Accessed 14 April 2020.

CHAPTER 15

Plastics and Social Responsibility

Susan Mossman

Plastics are now the most widely used materials group in the world. Global production of plastics has increased twentyfold since the 1960s, reaching around 360 million mt[1] in 2018 (Garside 2019). By 2050, these figures are expected to quadruple (Qualman 2017). The extent to which plastics materials are used, combined with many of the products being intended for single use, has given plastics a bad name but the fault lies with us rather than with the material itself. This chapter explores the scale of production, its environmental consequences, the human role in the creation of the plastics waste mountain, the value widely attached to the plastic things that form it, and ways of addressing the problem.

Plastics are contradictory materials, variously considered as useful and adaptable, but also often considered to be cheap and throwaway. This perception grew more prevalent in the 1950s with the rising use of plastics in various applications, including toys, consumer goods and the increasing proliferation of packaging. Another less common viewpoint is that modern plastics may be expensive and highly engineered materials designed to perform in challenging environments and to demanding standards, whether in high-tech applications such as aerospace, high-end sporting

S. Mossman (✉)
Science Museum, London, UK
e-mail: Susan.Mossman@ScienceMuseum.ac.uk

© The Author(s), under exclusive license to Springer Nature Switzerland AG 2020
S. Lambert (ed.), *Provocative Plastics*,
https://doi.org/10.1007/978-3-030-55882-6_15

equipment, electronics, or highly specific medical uses such as bioengineering. During the 2020 Coronavirus pandemic, plastics have become of vital importance for use in certain items, including personal protective equipment for medical and care staff and in ventilator components. Some commonplace perceptions of plastics actively misrepresent their value and discourage their sustainable use, reuse, recycling and disposal. Provoking a transformation in customary attitudes towards plastics and their perceived value may well help to cultivate a sense of social responsibility towards them.

Only in the last twenty years, have Britain and Western Europe become more attuned to the idea of social responsibility related to the manufacture, use and disposal of plastics. From the 1950s onwards, the main emphasis was on the modernity, convenience, practicality and cheapness of plastics compared with more traditional materials. Elsewhere attitudes have varied and sometimes 'elsewhere', notably the oceans, has increasingly become garbage dumps for those plastics that 'we' choose to throw away (Fig. 15.1).

Plastics' pollution in oceans has evoked widespread outrage in the media and public attitudes towards this blight on our natural environment

Fig. 15.1 Classic example of a catcher beach along the shores of Hawaii. (Photo: NOAA)

are hardening. Plastics waste is rising annually, mainly due to increasing consumption of 'single-use' plastics, including packaging and other mass-produced consumer items discarded after one use, and rarely recycled. The most vilified items have been plastic water bottles, packaging, and bags: recent research found 122 million plastic bags along 5000 km of European coastline (Environment and Energy Committee 2019, 20; SAPEA 2019).

REFLECTING BACK

Plastics are often considered to be a twentieth-century phenomenon, beginning with the first phenol formaldehyde-based synthetic plastic, Bakelite. However, plastics have their semi-synthetic roots in the mid-nineteenth century with vulcanised rubber and cellulose nitrate-based materials such as Parkesine. Even earlier are the natural plastics, materials that could be formed under heat and moulded into various forms, notably horn and tortoiseshell.

Throughout the development of plastics, their makers and users have varied in their approach to social responsibility. From 1877 onwards, the British Xylonite Company marketed their products as an alternative to the increasingly rare and expensive elephant ivory and tortoiseshell, revealed by their company logo showing an elephant and a tortoise. However, as early as the 1909 patenting of Bakelite by Leo Baekeland, likened to a Grand Duke and wizard (Kaufmann 1968), there seemed to be little thought given to the sustainability of this thermosetting material which is made from hazardous chemicals and has to be ground down to be recycled. By the 1920s, The Bakelite Corporation was promoting Bakelite as a magical material. The extravagant language continued. *The Bakelite story* (Mumford 1924) was so fulsome in its praise that even Baekeland thought that there was 'Something not quite right about it', distancing himself from the account (LHB 1924). In 1924, The Bakelite Corporation issued *A Romance of Industry*, a booklet aimed at female consumers, using imagery such as the genie lamp, again emphasising the magical theme. In 1937, The Bakelite Corporation's self-aggrandising film, *The Fourth Kingdom*, placed Bakelite on the same level as the animal, vegetable and mineral kingdoms, celebrating Bakelite's 'magical' qualities and as 'a material of a 1000 uses'.

Perhaps this highly optimistic approach to the new world of plastics informed the later thoughts of French philosopher Roland Barthes, who, in 1957, referred to plastics as a 'magical substance' capable of 'infinite transformation' (Barthes 1972). This contradicts the viewpoint of those modern environmentalists who have demonised plastics, recording plastics garbage littering land and sea with increasing frequency and passion. With up to 8 million mt of plastics waste being dumped in the oceans each year adding to the estimated 150 million mt already present, the modern perspective deserves very serious consideration as does the urgent need for steps to be taken to diminish these figures for ocean plastics waste (Ocean Conservancy 2020).

ACTION AGAINST PLASTICS WASTE

In 2018, UK Prime Minister Theresa May tweeted: 'In 2015 we introduced the 5p charge on plastic carrier bags, we now see 9 bn fewer bags being used. It's making a real difference. We want to do the same with single use plastics. Nobody who watched #BluePlanet2 will doubt the need for us to do something—and we will' (May 2018). Although there is debate as to how effective plastics bans really are. The UN estimates that at least one trillion plastic bags are made globally, so bag production is still a large use of a precious feedstock resource, namely oil (Parker 2019a).

In 2018, the UK's London *Evening Standard* newspaper campaigned to reduce or remove the use of disposable plastics straws (Prynn and Edwards 2018), and in 2019 the UK government confirmed a forthcoming ban on them (DEFRA 2019). Pressure is increasing to reduce the amount of single-use plastic bottles. Notably, the 2019 Glastonbury festival organisers encouraged visitors to bring their own refillable water containers and banned the sale of single-use plastic bottles (Herbert 2019; Marsh 2019). Some local UK councils are increasing the availability of drinking water fountains as are public venues such as airports.

Quantities of plastic-stemmed cotton-buds in plastics waste resulted in the cotton bud project and related legislation to ban plastics stems. Cotton buds were banned in Scotland in 2019 with EU plans to ban them in 2021.

Single-use PPE such as masks and gloves, which have become so essential during the 2020 Coronavirus pandemic, are now proving problematic when disposed of carelessly (Allison et al. 2020).

The use of microplastics has provoked debate. These plastic particles, smaller than 5 mm, are used in body scrubs, toothpastes and other cosmetic applications. Microplastics have been found to accumulate in the sea, where they are ingested by marine life. The UK banned the use of microplastics in 2018 (Carrington 2018).

There are also concerns about microfibres entering the environment and the animal kingdom through water routes: 'Tiny plastic particles washed off products such as synthetic clothes and car tyres could contribute up to 30% of the "plastic soup" polluting the world's oceans and … are a bigger source of marine plastic pollution than plastic waste' (IUCN 2017; Boucher and Friot 2017; Cole et al. 2011).

In the USA, Perfluoroalkyl and polyfluoroalkyl substances (PFAS) have been banned in food packaging as they have been found to be harmful, commencing with a ban in California on 1 January 2020, with other bans to follow (Vorst 2020; Safer States 2020; Ross 2019).

China noted the value of waste plastics and took steps to acquire and repurpose the plastics that Europe and North America threw away. By 2016 China was importing 45% of the world's plastics waste (c.10225 mt) (Brooks et al. 2018). Until the end of 2017, Britain shipped up to two-thirds of its waste plastics to China for recycling: about 500,000 tonnes a year (Taylor 2018a). However, Chinese processing facilities became overrun with vast quantities of contaminated materials, creating an environmental problem (Katz 2019). Thus, from 1 January 2018, China banned imported plastics waste (Hataway 2018). This ban is challenging for the EU and the USA and has greatly impacted on the UK's recycling industry (Bodkin 2018). In reaction, the UK and some other developed countries have begun to send their plastics waste to other Southeast Asian countries, such as Thailand and Malaysia. Consequently, some Chinese recyclers have seized the opportunity to open new processing plants in nearby countries (UNEP 2018). Longer-term, local solutions to deal with local waste would be preferable.

Future sustainability depends on increasing the awareness of individuals, organisations and government departments of what happens to their waste plastics. This will enhance social responsibility at all levels. Efforts made at every scale can help manifest a transformation in the way we look at our 'waste' material and see it from a different perspective. The future archaeologist (or garbologist) may excavate landfill piles of waste plastics, regarding their finds as treasure. Alternatively, plastics mountains may become a source of fuel or material to be recycled and turned into new plastics.

To help meet these challenges, the 2015 strategy document entitled 'A Vision for Europe's New Plastics Economy' promulgated: 'A smart, innovative and sustainable plastics industry, where design and production fully respects the needs of reuse, repair, and recycling, brings growth and jobs to Europe and helps cut EU's greenhouse gas emissions and dependence on imported fossil fuels' (EC 2018a, 2019b, 5). Certain individuals and companies have led the way with a more sustainable approach, whether in manufacturing clothing made from recycled PET bottles or furniture made from waste plastics packaging. MIT has converted discarded plastic soda bottles into solar bottle bulbs to help light thousands of homes of the poor in the Philippines. The 'bulb' is made from a one-litre soda bottle filled with purified water and bleach, inserted tightly into a house roof. The clear water disperses the sunlight and functions like an electric light bulb, costing only 2 or 3 dollars (Oshima 2011; Chen et al. 2014).

The integration of designers into initial product design at earlier stages of projects has contributed a more coherent approach to plastics manufacturing. Credit in the UK for supporting this methodology goes to the government-funded Knowledge Transfer Network. Since 2005, it has facilitated introducing designers to industrial and materials' manufacturers to grow mutual understanding, increase comprehension and collaboration, and develop this interdisciplinary community (KTN 2020). The KTN are partners with the Materials and Design Exchange, which publishes the Journal MaDE® (Materials and Design Exchange 2020).

PLASTICS WASTE AND THE GREAT PACIFIC GARBAGE PATCH

Nonetheless, the plastics waste mountain is forever growing. About 8.3 billion mt of plastics have been produced worldwide since the 1950s, reflecting the enormous increase in their use for packaging. Annual global plastics production reached almost 360 million mt in 2018 (PlasticsEurope 2019, 14) with until 2017, only 9 percent of these plastics being recycled (Sönnichsen 2020).

In 2018, the EU produced almost 62 million mt of plastics. Almost 39.9% of the 51.2 mt of plastics converted in Europe's manufacturing plants went into packaging; 19.8% went into building and construction, the second-largest application of plastics (PlasticsEurope 2019, 20). The recycling and recovery rate of packaging waste doubled from 2006 to 2018, with over half the EU countries recycling above 40% of their plastics packaging, although annually plastics waste is still increasing (PlasticsEurope 2019, 31). In 2018, Europe collected 29.1 million mt of total plastics waste, of which 32.5% was recycled, 42.6% went to energy recovery and 24.9% to landfill (PlasticsEurope 2019, 29). Packaging waste comprised 17.8 mt of this total: 7.5 (42%) mt were recycled, 7 (39.5%) mt went into energy recovery and 3.3 (18.5%) mt into landfill. In 2019, the EU reused less than 12% of its recycled materials so it is encouraging that 'Work is ongoing to review the essential requirements for packaging, which will aim at improving design for reuse and high-quality recycling of packaging materials' (EC 2019a, 1, 2).

Legislation has raised plastics packaging recycling levels. The EU has issued a series of packaging laws, key being the 1994 Directive 94/62/EC on packaging and packaging waste legislation. It focused on minimising the creation of packaging waste, promoting the reuse, recycling and recovery of energy from packaging and avoiding single use packaging. Subsequently EU Directive 2018/852 amended Directive 94/62/EC setting the minimum recycling target for plastics in packaging waste at 50% by 31 December 2025, rising to 55% by weight by 31 December 2030 (Packaging Waste Directive 2018).

Why is this legislation so important? The world's annual consumption of plastics materials increased from around 5 million tonnes in the 1950s to nearly 360 million mt in 2019 (Garside 2019). In Europe, in 2018, 24.9% of the plastics waste ended up in landfill, and contained 3.3% packaging. This is an improvement on EU 2006 figures when 7.2% of plastics packaging waste went to landfill, but globally the picture is not so optimistic. Worldwide, 40% of plastics waste goes to landfill with a plastics consumption of 45 kg per head (Wang 2020). It is estimated that, buried in landfill, a plastic bottle or bag will take up to 1000 years to decompose. Also degrading plastics may leak pollutants into the surrounding soil and water. Alternatively, carelessness may lead to plastics waste ending in the oceans which already contain an estimated 150 million mt of floating plastics debris, with quantities entering marine and bird life, proving hazardous to their health and sometimes even fatal. The icon of the British media, Sir David Attenborough, brought wider recognition to this issue in the 2017 *Blue Planet II* television series (Blue Planet II 2018). In response, on 22 November 2017, the British government undertook to penalise single-use plastics via tax measures. Former Chancellor of the Exchequer Philip Hammond stated: 'I want us to become a world leader in tackling the scourge of plastic' (Gabbatis 2017). The British Plastics Federation commented: 'plastic needs to be used responsibly and where it provides value—and ultimately recycled in all cases where possible … .We can all make a difference' (BPF 2020).

The UK annually generates nearly 5 million mt plastics, of which 2.4 million mt is packaging (WRAP 2019, 2; House of Commons 2020).

In 2019, authorities collected approximately 550 kT^2 of packaging waste, 10% more than 2013–2014. Only two types of plastics are consistently recycled: polyethylene terephthalate (PET) and high-density polyethylene (HDPE). The UK Plastics Pact, launched in April 2018, aims to create a circular economy for plastics packaging with 2025 goals for plastics packaging to be 100% reusable, recyclable or compostable and the elimination of 'problematic or unnecessary single use packaging … through redesign, innovation or alternative (reuse) methods' (WRAP 2018).

Some European countries still use landfill extensively for waste plastics. In 2011, the European plastics industry launched the initiative 'zero plastics to landfill', to reduce the amount of post-consumer plastics waste sent to landfills to zero by 2020. This target has not been met. Seven EU member states have introduced landfill bans on plastics waste although, following their bans, incineration rates increased in the Netherlands, Germany and Denmark (Zero Waste Europe 2015). By 2016, the EU plus Norway and Switzerland had collected 25.1 million mt of post-consumer plastics waste, a recovery rate of 72.7%. The highest recycling and recovery rates were for plastics packaging; and, in a first, more plastics waste was recycled than landfilled. In 2018, the EU sent 7.2 (24.9% of total) million mt of plastics waste to landfill (PlasticsEurope 2019, 28). The target date of 2020 'zero plastics to landfill' was modified to 2025 as was the mandatory separate collection of all packaging from residual waste, both by the year 2025 (Bioplastics 2016).

However, there is an ongoing problem. In 2011, 61.5% of UK beach litter was plastics and, in 2015, the Marine Conservation Society recorded 160 plastic bottles per mile of UK coastline they had cleaned (Marine Conservation Society 2013, 8; Daniels 2016, 7). In 2018, a million volunteer-strong initiative to clean up beaches, involving 120 countries, found plastics food packaging was the most common form of rubbish (Parker 2019b).

Globally, much of the plastics found in the oceans travels there from ten rivers, running largely through areas with no rubbish collection facilities nor programmes to turn waste to advantage, including the Yangtze, Mekong and Indus rivers (MoDiP 2020). Captain Charles Moore observed in 2003, while sailing back to Long Beach, California: '… we decided to take a shortcut through the gyre, which few seafarers ever cross. … Yet as I gazed … at the surface of what ought to have been a pristine ocean, I was confronted, as far as the eye could see, with the sight of plastic' (Moore 2003).

About 54% of the Great Pacific Garbage Patch (GPGP) debris originates from land-based activities in North America and Asia. The rubbish takes about six years to reach the GPGP from the coast of North America and a year from Japan and other Asian countries (National Geographic 2020). Most of the remaining garbage is fishing gear such as ropes and nets dumped by sea vessels (National Geographic 2019). In 2018, the GPGP was estimated to contain nearly 80,000 tonnes of plastics in 1.6 million square kilometres (Gabbatis 2018). By 30 April 2020, it had reached almost 2 million square kilometres, growing at a rate of almost 350 square kilometres daily (The World Counts 2020). It is the largest of five such subtropical garbage patches (gyres) found worldwide (The 5 Gyres Institute 2020). Due to their growing use in increasing numbers of goods, plastics make up the largest proportion of ocean waste. That many plastics products do not biodegrade in sea water but instead break down into smaller pieces is also problematic.

NEW APPROACHES, FUTURE STRATEGIES AND TECHNOLOGIES

An alternative to recycling or burying plastics is converting them into crude oil or other types of liquid fuel through pyrolysis, using high temperature techniques. American company Agilyx produces processing units that can be placed where the plastics waste is collected, for example, in municipal waste facilities. This system transforms ground, unsorted plastics into synthetic crude oil which can then be refined into ultra-low sulphur diesel, gasoline, or jet fuel and into synthetic lubricants and greases. Some of these products can then be turned back into plastics. In 2018, Agilyx adapted their pyrolysis technology to produce a naphtha base-stock which may enable the production of olefins from recycled materials. One potential product might be feedstock for polypropylene, the plastic most in demand worldwide (ICIS 2018).

Waste plastics can also be burned in waste-to-energy (WTE) plants which recover energy and reduce the quantity of plastics in landfills. Inorganic carbon in plastics does not decompose through anaerobic digestion processes, so there are no emissions if it is landfilled (Lee et al. 2017, 340). WTE processes using non-recycled plastics, release carbon into the atmosphere during fuel production and combustion processes as in other WTE technologies. Such a WTE plant is situated in Baltimore, USA

(Wheelbrator Technologies 2020). These plants produce electricity and heat in boilers, which are designed for complete combustion and are reported to produce electricity 'with less environmental impact than almost any other source of electricity' (Sustainable waste solutions 2020). However, they are often perceived locally as blights on the landscape and accused incorrectly of giving off noxious fumes. Better design would improve the first and the 'noxious fumes' misperception could be eradicated by better public education.

A 2009 report on converting plastics waste into a resource described the production of gaseous fuels by using high temperatures to decompose plastics waste, obtaining solid fuel, from a mixture of waste plastics, wood and paper (UNEP 2009). The forecast is that energy recovery will become the second-largest plastics waste destination in the future. The main impetus for this in Europe is the Waste Framework Directive revisions stipulating a limit of 10% waste going to landfill by 2030 (EC 2019b). In 2018, Europe sent 42.6% of its plastic waste to energy recovery.

It is also important to develop new ways of making plastics not dependent on fossil fuels. Innovative research is ongoing into plastics based on natural materials including cellulose, algae and bacteria, a field known as bioplastics or biopolymers. Bioplastics are made from renewable feedstock, some being biodegradable or compostable. An early example is ICI's *Biopol*, first commercialised in 1990, made from polyhydroxybutyrate (PHB), obtained from the bacterium *Alcaligenes eutrophus*, and degrading to form carbon dioxide and water (ICIS 1991; New Scientist 1990). Some other bioplastics are based on foodstuffs such as sugar, starch and corn, including polyhydroxyalkanoates (PHA), polylactic acids (PLA), thermoplastic starches (TPS) and feedstocks from vegetable oils. However, less than 40% of these bioplastics are designed to be biodegradable (Vorst 2020; Barrett 2019) and often need optimum conditions to degrade completely. When they are combined with other plastics components to reduce cost and/or improve properties, the degradation may only be partial; PLA bioplastics do not degrade in sea water (Barrett 2019). Again, with global population increasing, using food stocks to make plastics needs very careful consideration to ensure that human food requirements are not sacrificed in return for yet more consumer goods.

A further active field is reconsideration of semi-synthetic plastics based on cellulose but using more sustainable production processes, such as less toxic solvents. Tencel© is a noteworthy case, developed to be a more environmentally friendly form of viscose rayon by Courtaulds in the late 1970s (Blanc 2016, 217).

Developers of ever more complex plastics and designers of future plastics products will increasingly have to consider the entire life cycle of their material and product design, adopting a cradle-to-cradle approach, rather than cradle to landfill or ocean grave. It is vital that industrial designers work collaboratively with other disciplines including polymer scientists, environmental and sustainability engineers and specialists, to ensure that their new designs integrate the principles of sustainability and social responsibility. A MoDiP exhibition *Encore* displayed examples of the many ways in which plastics can be recycled, reworked and repurposed (MoDiP 2009).

New ways of making and treating plastics waste are in development, a vital aspect being new methods of making packaging. When possible, it is desirable to change from using complex multi-layered plastics that, when recycled, produce a very low quality and low cash value recyclate to packaging made of one or two closely related plastics that can be turned into a good quality reusable material. Research is ongoing to blend recycled polyethylene/polypropylene with other waste fillers to produce a higher quality recyclate (Vorst 2020). Others are piloting upcycling technologies to transform waste polyethylene into high quality monomers for onward use in high quality products (Biocellection 2020). Developments continue on improving compostable packaging, for example, cellulose-based food packaging, applications including coffee bags and capsules (Natureflex 2020).

Although efforts need to be made to reduce plastics packaging use, this is not as simple as it sounds as the downside of returning to a plastics-free packaging world is that more food would go to waste. The Netherlands' initiative is the 'no plastics' shopping aisle but this is misleading as goods are wrapped with compostable bioplastics (Taylor 2018b). The shopper who unpacks their purchases at the counter and returns the waste packaging to the retailer can also contribute personally to a shift in attitudes.

Reduction of the GPGP requires major global initiatives. 'Hoovering' the sea is a possible route, although this might endanger marine life. It is clear from the evidence that using alternatives to damaging and toxic plastics additives, such as bisphenols that are poorly soluble in water and toxic to marine life, should be sought. Teams are seeking different ways to break down plastics packaging waste. In 2016, a study showed how specialised bacteria might be used to biodegrade PET bottles (Yoshida et al. 2016).

The Ocean Cleanup team aim to remove 90% of ocean plastics by 2040 using a floater system that captures plastics, using natural forces, and is slowed down by a sea anchor. Once full, the system will be emptied by a sea-going vessel (The Ocean Cleanup 2020).

However, most importantly, what is essential is a change of mindset. Mass consumption of single-use goods that grew rapidly from the 1950s onwards, to which the cheapness, easy availability, malleability and practicality of plastics contributed, must be largely abandoned. Perhaps a return to an older perception of plastics as being valuable and sometimes even precious materials that should be used with discrimination and care would provoke a major transformation. The early cellulose-nitrate dressing table sets made by companies such as Xylonite (British Xylonite 1927) and the iconic vintage EKCO Bakelite radios are cases in point (EKCO 1939).

Artists' Responses to Plastics Sustainability

Artists are using plastics to comment on sustainability and suggest their value. Two British examples are Michelle Brand and Richard Sowa. Brand turns plastic bottles into decorative items (Brand 2020) and Sowa made an island from 150,000 recycled plastic bottles (Carroll and Roberts 2014). The Ghanaian artist Serge Attukwei Clottey explores his country's culture of reuse, turning plastic jugs, locally known as Kufuor gallons, into works of art (Jansen 2016).

Joshua Sofaer's *The Rubbish Collection* exhibition held at London's Science Museum in summer 2014 continued a series of thought-provoking displays and events related to the museum's 2010 *Atmosphere* gallery and associated five-year *Climate Changing* programme. Playing with the recognisable role of the museum in collecting, sorting and displaying precious objects, and using them to tell stories, he explored that which the museum throws away institutionally and additionally as individual staff and visitors. Firstly, project staff, volunteers and museum visitors sorted and documented the Museum's rubbish for 30 days. Secondly, the equivalent amounts of transformed materials produced from the rubbish were brought back into the museum and displayed beside items retained from the original rubbish. The preparatory stages stimulated conversations within the museum and the subsequent self-reflection influenced decisions the institution made concerning future sustainability and climate change. The installation itself clearly presented that which the museum itself produced in the form of waste and exposed the value of these overlooked materials, in aesthetic and monetary terms. The concept was surprising and had the potential to shock as Sofaer brought his audience and the museum's visitors face to face with their daily consumption and waste of resources.

The team had predicted that around 28 tonnes of rubbish would be thrown out. Over 18 tonnes of materials were brought back to the gallery for the exhibition's second phase, including 7.4 tonnes of paper and card reels, 2.4 tonnes of bottom ash aggregate, 2.3 tonnes of glass sand, 1.4 tonnes of wood, one tonne of fertiliser, 698 kilograms of steel, 650 litres of dehydrated sewage sludge, 291 breezeblocks made from air pollution control residue and nearly one tonne of various recycled plastics. Items retained included an astonishing quantity of plastics cutlery.

When the display closed, the materials, including the plastic disposable pellets, were returned to the companies that had lent them and were then turned into new products. Electrical goods were sent to specialist recycling companies to separate any reusable parts and recycle what could be salvaged. Most of the items retained from the rubbish bags would have originally gone to incineration if they had not been selected. These items were recycled wherever possible although some were incinerated.

At the recycling facility, the rubbish was separated into different recycling streams. Magnets separated the ferrous from the non-ferrous metals. Infra-red technology identified and sorted the remaining rubbish into paper, card, glass and several types of plastics. Rubbish that could not be recycled, or pieces that were too small to be captured, were taken elsewhere for incineration. Nothing went to landfill. The materials were then baled as raw materials for resale to companies who took on the next stage of processing.

The recycling and recovery processes were different for each material. There is always a loss of some material, which cannot be recovered or usefully re-used. With plastics and glass, the loss came from paper labels and glues that were soaked off in the washing process, forming a sticky substance which was sent for incineration. A future is envisaged where these materials will either be captured for future use, or the waste will be designed out altogether. By using plastics labels instead of paper, the material could be more easily collected and recycled to make new products such as plastic bags.

The recycling of this rubbish produced some useful and valuable products. One discovery was that much recyclable material was incinerated. Those materials retain much more value when they are recycled so by continuing to improve and refine institutional recycling systems, and through new initiatives like separating food waste, general waste could be decreased further in the future.

Joshua Sofaer concluded: 'the very thing that this project has relied on, that people throw stuff away, is also the thing we want to reduce. Let's work towards a time when a project like this is unnecessary or even impossible. Disposal is the last resort' (Sofaer 2014).

Other more recent, linked, museum initiatives include the British Museum's exhibition *Disposable? Rubbish and Us* and the Swedish Röhsska Museet's exhibition *Ocean Plastics*, both in 2019. The first explored how humans have interacted with rubbish over the millennia including producing single-use items over time and commented on the modern 'unprecedented levels of waste' (British Museum 2019). The second presented a selection of conceptual design projects underpinned by an interdisciplinary approach and a belief in design's ability to contribute to the solution rather than the problem. Themes included cleaning the oceans, plastics recycling and bioplastics (Röhsska 2020).

CONCLUSION

Bans of plastic bags and single-use plastics seize headlines but are only the start of what needs to be done to deal with the much bigger issue. Karl-H. Foerster, the former executive director of PlasticsEurope, stated that: 'To protect our environment effectively, we need to educate citizens so they understand that plastics are too valuable to be thrown away' (Foerster 2014). This perception that educating the public is key to a more balanced and responsible view of plastics is something that may become more evident as plastics waste increasingly becomes a rich resource for future reuse. The increasing plastics waste in the oceans provokes extreme concern. Behaviours must change, on personal, commercial, industrial and governmental levels, so that the casual littering of seashores and disposal of rubbish into the sea is reduced and ultimately prevented. International collaboration will be essential to support research into ways of designing disposal mechanisms for plastics waste that create value rather than devastation for the environment and to make recycling and energy recovery more effective. One country's waste should not become another's burden.

Modifying people's behaviour can be achieved through legislation, education, or by example. However, legislating can be a clumsy, expensive, and lengthy process and may cause resentment. Making disposing of plastics waste in a responsible way both easy and affordable, and irresponsible practices expensive and a matter for public shame, would be helpful. Education in schools and influencing behaviours from a young age is perhaps the most effective route. However, experience from developing-related exhibitions at the Science Museum in London, and elsewhere, has shown that it is much more effective to engage positively with the public rather than to preach to them.

NOTES

1. mt is abbreviation for metric tons.
2. Kt is abbreviation of kiloton (1000 tons).

BIBLIOGRAPHY

Allison, Ayse Lisa, Esther Ambrose-Dempster, et al. 2020. *The Environmental Dangers of Employing Single-Use Face Masks as Part of a COVID-19 Exit Strategy.* UCL Plastic Waste Innovation Hub, University College London, London, UK. London: UCL Press (Preprint). May 2020. DOI: 10.14324/111.444/000031.v1. Accessed 20 August 2020.

Barrett, Axel. 2019. Finding Solutions to Prevent Harmful Plastic Waste Is Far from Simple. *The Guardian,* July 8. https://bioplasticsnews.com/2019/07/08/the-guardian-writes-about-bioplastics/. Accessed 16 May 2020.

Barthes, Roland. 1972. *Mythologies.* Trans. Annette Lavers. London: Paladin. (Original French Edition 1957).

Biocellection. 2020. Upcycling Plastics on a Molecular Level. https://www.biocellection.com/. Accessed 16 May 2020.

Bioplastics Magazine.com. 2016. Zero Plastics to Landfill by 2025, and More. October 3. https://www.bioplasticsmagazine.com/en/news/meldungen/20161003-zero-plastics-to-landfill-by-2025-and-more.php. Accessed 29 April 2020.

Blanc, Paul David. 2016. *Fake Silk: The Lethal History of Viscose Rayon.* New Haven, VT: Yale University Press.

Blue Planet II. 2018. http://www.bbc.co.uk/programmes/p04tjbtx. Accessed 6 April 2018.

Bodkin, H. 2018. Toxic Plastic to be 'Burned in Britain' Due to China Import Ban. *The Telegraph,* January 1. https://www.telegraph.co.uk/news/2018/01/01/toxic-plastic-burned-britain-due-china-import-ban/. Accessed 28 March 2020.

Boucher, Julien, and Damion Friot. 2017. *Primary Microplastics in the Oceans: A Global Evaluation of Sources.* Gland, Switzerland: IUCN.

BPF. 2020. British Plastics Federation Responds to BBC's War on Plastic. https://www.bpf.co.uk/article/british-plastics-federation-responds-to-bbc-war-on-plastic-1470.aspx. Accessed 1 May 2020.

Brand, Michelle. 2020. Environmental Design. https://www.michellebrand.co.uk. Accessed 1 May 2020.

British Museum. 2019. Past Exhibition, Disposable? Rubbish and Us. https://www.britishmuseum.org/exhibitions/disposable-rubbish-and-us.   Accessed 1 May 2020.

British Xylonite. 1927. 1927 Catalogue, Hale End, British Xylonite.

Brooks, Amy L., Shunli Wang, and Jenna R. Jambeck. 2018. The Chinese Import Ban and Its Impact on Global Plastic Waste Trade. *Science Advances* 4 (6): eaat0131.   https://advances.sciencemag.org/content/4/6/eaat0131/tab-. Accessed 29 April 2020.

Carrington, Damian. 2018. Plastic Microbeads Ban Enters Force in UK. *The Guardian*, January 9. https://www.theguardian.com/environment/2018/jan/09/plastic-microbeads-ban-enters-force-in-uk. Accessed 23 March 2018.

Carroll, Michael, and Gareth Roberts. 2014. Meet British 'Robinson Crusoe' Who Lives with Supermodel on Island He Made Out of 150,000 Recycled Bottles. *Daily Mirror*, November 7.

Chen, Wang, et al. 2014. Critical View on Lighting Through a Solar Bottle Bulb. *Building Research Journal* 61 (2): 115–128. https://www.researchgate.net/publication/273912152_Critical_View_on_Daylighting_Through_Solar_Bottle_Bulb. Accessed 12 May 2020.

Cole, Matthew, Pennie Lindeque, Claudia Halsband, and Tamar S. Galloway. 2011. Microplastics as Contaminants in the Marine Environment: A Review. *Marine Pollution Bulletin* 62 (12): 2588–2597.

Daniels, Natalie. 2016. Cleaning Up the Problem. *Materials World*, 5, May 2016: 7.

DEFRA. 2019. Press Release: Gove Takes Action to Ban Plastic Straws, Stirrers, and Cotton Buds. Department for Environment, Food & Rural Affairs. May 22.   https://www.gov.uk/government/news/gove-takes-action-to-ban-plastic-straws-stirrers-and-cotton-buds. Accessed 29 April 2020.

EC. 2018a. Communication from the Commission to the European Parliament, the Council, the European Economic and Social Committee and the Committee of the Regions: A European Strategy for Plastics in a Circular Economy, European Commission, Brussels, 16.1.2018 COM (2018) 28 final, 2018:3. https://eur-lex.europa.eu/legal-content/EN/TXT/?uri=COM:2018:28:FIN. Accessed 29 April 2020.

———. 2018b. Plastic Waste: A European Strategy to Protect the Planet, Defend Our Citizens and Empower Our Industries, Press Release. European Commission. January 16. https://ec.europa.eu/commission/presscorner/detail/en/IP_18_5. Accessed 29 April 2020.

———. 2019a. Report from the Commission to the European Parliament, the Council, the European Economic and Social Committee and the Committee of the Regions on the implementation of the Circular Economy Action Plan. European Commission. https://ec.europa.eu/commission/sites/beta-political/files/report_implementation_circular_economy_action_plan.pdf. Accessed 1 May 2020.

———. 2019b. Review of Waste Policy and Legislation, updated August 7. European Commission. https://ec.europa.eu/environment/waste/target_review.htm. Accessed 1 May 2020.

EKCO. 1939. EKCO Ltd Catalogue 1939/40, Southend Museum Collection.

Environment and Energy Committee. 2019. Solutions to Plastics in the Ocean – The Baltic and Beyond. *Swedish Academy of Sciences*, (Symposium Proceedings). https://s3.eu-de.cloud-object-storage.appdomain.cloud/kva-image-pdf/2020/03/Symposium_rapport_final_webb.pdf. Accessed 29 April 2020.

Foerster, Karl-H. 2014. PM+: Landfill Ban on Recyclable Materials Makes 'Economic and Environmental' Sense. *The Parliament Magazine*, December 8. https://www.theparliamentmagazine.eu/articles/sponsored_article/pm-landfill-ban-recyclable-materials-makes-economic-and-environmental. Accessed 1 May 2020.

Gabbatis, Josh. 2017. Budget 2017: Philip Hammond Reveals Plans for a Plastic Tax and Announces Clean Air Measures. *Independent Online*, November 22. https://www.independent.co.uk/environment/budget-2017-air-pollution-philip-hammond-plastic-tax-clean-measures-statement-environment-a8069946.html. Accessed 5 April 2018.

———. 2018. 'Great Pacific Garbage Patch' 16 Times Bigger than Previously Thought, Say Scientists. *Independent*, March 23. https://www.independent.co.uk/environment/great-pacific-garbage-patch-plastic-pollution-oceans-environment-fish-a8269951.html. Accessed 29 April 2020.

Garside, M. 2019. Global Plastic Production Statistics. *Statista.com*, November 8. https://www.statista.com/282732/global-production-of-plastics-since-1950. Accessed 7 April 2020.

Hataway, James. 2018. Scientists Calculate Impact of China's Ban on Plastic Waste Imports. *UGA Today*, June 20. https://news.uga.edu/topics/science-technology/. Accessed 29 April 2020.

Herbert, Tom. 2019. Glastonbury Plastic Bottles Ban 2019: Has Glasto Gone Plastic Free? Which Single Use Plastics Has the Festival Banned? *Evening Standard*, June 25. https://www.standard.co.uk/futurelondon/theplastic-freeproject/glastonbury-plastic-bottles-ban-single-use-plastic-free-festival-a4175076.html. Accessed 7 January2020.

House of Commons. 2020. Plastics Waste (Briefing Paper). March 31. https://commonslibrary.parliament.uk/research-briefings/cbp-8515/. Accessed 1 May 2020.

ICIS. 1991. ICI Reduces Cost, Ups Capacity for Biopol. September 22. https://www.icis.com/explore/resources/news/1991/09/23/25670/ici-reduces-cost-ups-capacity-for-biopol/. Accessed 29 April 2020.

———. 2018. ICIS Insight: US Agilyx Takes Big Step in Producing Naphtha, Propylene from Waste Plastics. September 27. https://www.agilyx.com/newsroom/icisinsight-us-agilyx-takes-big-step-producing-naphtha-propylene-waste-plastics. Accessed 1 May 2020.

IUCN. 2017. Invisible Plastic Particles from Textiles and Tyres a Major Source of Ocean Pollution – IUCN Study. February 22. https://www.iucn.org/news/secretariat/201702/invisible-plastic-particles-textiles-and-tyres-major-source-ocean-pollution-%E2%80%93-iucn-study. Accessed 28.3.2018.

Jansen, Charlotte. 2016. The Ghanaian Turning Thousands of Discarded Plastic Bottles into Art. *The Guardian*, March 31. Last modified on 22 February 2018. http://www.theguardian.com/world/2016/mar/31/ghana-art-serge-attukwei-clottey-pastic-pollution-exhibition?CMP=Share_iOSApp_Other. Accessed 31 March 2016. Revisited 29 April 2020.

Katz, Cheryl. 2019. Piling Up: How China's Ban on Importing Waste Has Stalled Global Recycling, Yale Environment 360. *Yale School of Forestry and Environmental Studies*. March 7. https://e360.yale.edu/features/piling-up-how-chinas-ban-on-importing-waste-has-stalled-global-recycling. Accessed 29 April 2020.

Kaufmann, C. 1968. *Grand Duke, Wizard and Bohemian: A Biographical Profile of Leo Hendrik Baekeland (1863–1944)*. MA thesis, University of Delaware.

KTN. 2020. https://ktn-uk.co.uk/about. Accessed 7 January 2020.

Lee, Uisung, Jeongwoo Han, and Michael Wang. 2017. Evaluation of Landfill Gas Emissions from Municipal Solid Waste Landfills for the Life-Cycle Analysis of Waste-to-Energy Pathways. *Journal of Cleaner Production* 166: 335–342.

LHB. 1924. Correspondence Between Arthur Little and Leo Baekeland: Little to Baekeland. October 25, 1924 and Baekeland to Little, October 27, 1924. Leo Baekeland Archive, Smithsonian Institution, LHB.

Marine Conservation Society. 2013. Marine Plastics Pollution Policy and Position Statement. V1, March. https://view.officeapps.live.com/op/view.aspx?src=https%3A%2F%2Fwww.mcsuk.org%2Fdownloads%2Fpollution%2Fpbf%2FMCS_Marine_Plastics_position_paper.doc. Accessed 29 April 2020.

Marsh, Sarah. 2019. Glastonbury Festival Bans Plastic Bottles. *The Guardian*, February 27. https://www.theguardian.com/music/2019/feb/27/glastonbury-festival-bans-plastic-bottles. Accessed 7 January 2020.

Materials and Design Exchange. 2020. https://made.partners/about. Accessed 7 January 2020.

May, Theresa. 2018. Tweet: @Theresa May 7 January, 2016. Accessed 29 April 2020.

MoDiP. 2009. *Encore*. http://www.modip.ac.uk/exhibitions/encore. Accessed 6 April 2018.

———. 2020. Case Study – Water Bottle. https://www.modip.ac.uk/plastics/plastics-and-environment/case-study-water-bottle. Accessed 29 April 2020.

Moore, Charles. 2003. Trashed: Across the Pacific Ocean, Plastics, Plastics, Everywhere. *Natural History Magazine*, November. http://www.naturalhistorymag.com/htmlsite/master.html?http://www.naturalhistorymag.com/htmlsite/1103/1103_feature.html. Accessed 5 September 2015.

Mumford, John. 1924. *The Story of Bakelite*. New York: Robert L. Stillson Co.

National Geographic. 2019. *National Geographic*. July. https://www.nationalgeographic.org/article/great-pacific-garbage-patch-isnt-what-you-think/. Accessed 29 April 2020.

———. 2020. Great Pacific Garbage. *National Geographic*. http://nationalgeographic.org/encyclopedia/great-pacific-garbage-patch/. Accessed 29 April 2020.

Natureflex. 2020. https://www.natureflex.com/uk/. Accessed 16 May.

New Scientist. 1990. Technology: Biodegradable Plastic Hits the Production Line. *New Scientist*, May 5. https://www.newscientist.com/article/mg12617154-000-technology-biodegradable-plastic-hits-the-production-line/. Accessed 30 April 2020.

Ocean Conservancy. 2020. https://oceanconservancy.org/trash-free-seas/plastics-in-the-ocean/. Accessed 20 April 2020.

Oshima, Kotoe. 2011. Plastic Bottles Light Up Lives. *CNN*, August 30. www.cnn.com/2011/WORLD/asiapcf/08/30/eco.philippines.bottle. Accessed 3 October 2015.

Packaging Waste Directive. 2018. Directive (EU) 2018/852 of the European Parliament and the Council of 30 May 2018 Amending Directive 94/62/EC on Packaging and Packaging Waste. https://eur-lex.europa.eu/legal-content/EN/TXT/PDF/?uri=CELEX:32018L0852. Accessed 29 April 2020.

Parker, Laura. 2019a. Plastic Bag Bans Are Spreading. But Are They Truly Effective? *National Geographic*, April 17. https://www.nationalgeographic.com/environment/2019/04/plastic-bag-bans-kenya-to-us-reduce-pollution/. Accessed 12 May 2020.

———. 2019b. Plastic Food Packaging Was Most Common Beach Trash in 2018. *National Geographic*, September 3. https://www.nationalgeographic.com/environment/2019/09/plastic-food-packaging-top-trash-global-beach-cleanup-2018/. Accessed 29 April 2020.

PlasticsEurope. 2019. Plastics – The Facts 2019. https://www.plasticseurope.org/en/resources/publications/1804-plastics-facts-2019. Accessed 29 April 2020.

Prynn, Jonathan, and Lizzie Edwards. 2018. The Last Straw: The Evening Standard Launches a New Campaign to Eradicate Plastic Straws from London's Streets. *Evening Standard*, January 15. https://www.standard.co.uk/news/uk/the-last-straw-the-evening-standard-launches-a-new-campaign-to-eradicate-plastic-straws-from-londons-a3740101.html. Accessed 29 April 2020.

Qualman, Darrin. 2017. Global Plastics Production, 1917–2050. December 17. https://www.darrinqualman.com/global-plastics-production/. Accessed 16 April 2020.

Röhsska. 2020. Ocean Plastics. https://rohsska.se/en/exhibitions/ocean-plastics/. Accessed 20 April 2020.

Ross, Rachel. 2019. What Are PFAS? *Live Science*, April 30. https://www.livescience.com/65364-pfas.html. Accessed 16 May 2020.

Safer States. 2020. PFAS. *Safer States*. https://www.saferstates.com/toxic-chemicals/pfas/. Accessed 16 May 2020.

SAPEA. 2019. A Scientific Perspective on Microplastics in Nature and Society. Evidence Review Report No. 4. Berlin: SAPEA. https://www.sapea.info/wp-content/uploads/report.pdf. Accessed 1 May 2020.

Sofaer, Joshua. 2014. The Rubbish Collection. http://www.joshuasofaer.com/2014/10/the-rubbish-collection. Accessed 20 April 2020.

Sönnichsen, N. 2020. Plastic Waste in the UK – Statistics & Facts. *Statista*, February 10. https://www.statista.com/topics/4918/plastic-waste-in-the-united-kingdom-uk/#dossierSummary__chapter1. Accessed 1 May 2020.

Sustainable Waste Solutions. 2020. http://landfillfree.com/energy-from-waste.php. Accessed 1 May 2020.

Taylor, Matthew. 2018a. Rubbish Already Building Up at UK Recycling Plants Due to China Import Ban. *The Guardian*, January 2. https://www.theguardian.com/environment/2018/jan/02/rubbish-already-building-up-at-uk-recycling-plants-due-to-china-import-ban. Accessed 29 April 2018.

———. 2018b. World's First Plastic-Free Aisle Opens in Netherlands Supermarket. *The Guardian*, February 28. https://www.theguardian.com/environment/2018/feb/28/worlds-first-plastic-free-aisle-opens-in-netherlands-supermarket. Accessed 28 January 2018.

The 5 Gyres Institute. 2020. https://www.5gyres.org/science. Accessed 29 April 2020.

The Ocean Cleanup. 2020. https://theoceancleanup.com/oceans/. Accessed 29 April 2020.

The World Counts. https://www.theworldcounts.com/challenges/planet-earth/waste/great-pacific-garbage-patch-size. Accessed 4 May 2020.

UNEP. 2009. Converting Waste Plastics into a Resource, Assessment Guidelines (Revised Guidelines), United Nations Environmental Programme Division of Technology, Industry and Economics. Osaka/Shiga: International Environmental Technology Centre. https://wedocs.unep.org/bitstream/handle/20.500.11822/8618/WastePlasticsEST_AssessmentGuidelines.pdf. Accessed 29 April 2020.

———. 2018. China's Trash Ban Lifts Lid on Global Recycling Woes But also Offers Opportunity. United Nations Environment Programme. July 6. https://www.unenvironment.org/news-and-stories/story/chinas-trash-ban-lifts-lid-global-recycling-woes-also-offers-opportunity. Accessed 2 May 2020.

Vorst, Keith. 2020. Understanding the What, How & Why of Packaging. Paper, Online Conference, *Polymers in Food Packaging*, May 15.

Wang, T. 2020. Plastic Waste Worldwide – Statistics & Facts. *Statista*, March 10. https://www.statista.com/topics/5401/global-plastic-waste/#dossierSummary__chapter1. Accessed 1 May 2020.

Wheelbrator Technologies. 2020. https://www.wtienergy.com/plant-locations/energy-from-waste/wheelabrator-baltimore. Accessed 1 May 2020.

WRAP. 2018. The Plastics Pact. https://www.wrap.org.uk/content/the-uk-plastics-pact. Accessed 1 May 2020.

———. 2019. Plastics Market Situation Report 2019. https://www.wrap.org.uk/sites/files/wrap/WRAP_Plastics_market_situation_report.pdf. Accessed 1 May 2020.

Yoshida, Shosuke, et al. 2016. A Bacterium that Degrades and Assimilates Poly(ethylene terephthalate). *Science* 351 (6278): 1196–1199.

Zero Waste Europe. 2015. Press Release, Landfill Ban a False Path to a Circular-Economy, 09/11/2015. https://zerowasteeurope.eu/2015/11/press-release-landfill-ban-a-false-path-to-a-circular-economy/. Accessed 1 May 2020.

INDEX[1]

[1] Note: Page numbers followed by 'n' refer to notes.

© The Author(s), under exclusive license to Springer Nature Switzerland AG 2020
S. Lambert (ed.), *Provocative Plastics*,
https://doi.org/10.1007/978-3-030-55882-6

Printed by Books on Demand, Germany